# Worlds Fantastic, Worlds Familiar

## A Guided Tour of the Solar System

Join Bonnie J. Buratti, a leading planetary astronomer, on this personal tour of NASA's latest discoveries. Moving through the Solar System from Mercury, Venus, Mars, past comets and asteroids, and the moons of the giant planets, to Pluto, and on to exoplanets, she gives vivid descriptions of landforms that are similar to those found on Earth but that are more fantastic. Sulfur-rich volcanoes and lakes on Io, active gullies on Mars, huge ice plumes and tar-like deposits on the moons of Saturn, hydrocarbon rivers and lakes on Titan, and nitrogen glaciers on Pluto are just some of the marvels that await you. Discover what it's like to be involved in a major scientific enterprise, with all its pitfalls and excitement, from the perspective of a female scientist. This engaging account of modern space exploration is written for non-specialist readers, from students in high school to enthusiasts of all ages beyond.

BONNIE J. BURATTI is a Senior Research Scientist at NASA's Jet Propulsion Laboratory, California Institute of Technology. With expertise on icy moons, comets, and asteroids, she holds degrees from MIT and Cornell University. She is currently serving on the *Cassini* and *New Horizons* science teams, and is the US Project Scientist for *Rosetta*. She is a past Chair of the Division for Planetary Sciences of the American Astronomical Society, and advises NASA. The author of over two hundred scientific papers, Buratti was awarded the NASA Exceptional Achievement Medal and the International Astronomical Union named asteroid 90502 "Buratti" in recognition of her work.

# Worlds Fantastic, Worlds Familiar

## A Guided Tour of the Solar System

BONNIE J. BURATTI

*Jet Propulsion Laboratory,*
*California Institute of Technology*

CAMBRIDGE
UNIVERSITY PRESS

# CAMBRIDGE
UNIVERSITY PRESS

University Printing House, Cambridge CB2 8BS, United Kingdom

One Liberty Plaza, 20th Floor, New York, NY 10006, USA

477 Williamstown Road, Port Melbourne, VIC 3207, Australia

4843/24, 2nd Floor, Ansari Road, Daryaganj, Delhi - 110002, India

79 Anson Road, #06-04/06, Singapore 079906

Cambridge University Press is part of the University of Cambridge.

It furthers the University's mission by disseminating knowledge in the pursuit of education, learning and research at the highest international levels of excellence.

www.cambridge.org
Information on this title: www.cambridge.org/9781107152748

First published 2017

Printed in the United Kingdom by TJ International Ltd. Padstow Cornwall

*A catalogue record for this publication is available from the British Library.*

*Library of Congress Cataloging in Publication Data*
Names: Buratti, Bonnie Jean, 1952–
Title: Worlds fantastic, worlds familiar / Bonnie J. Buratti, Jet Propulsion Laboratory, California Institute of Technology.
Description: Cambridge : Cambridge University Press, 2017. | Includes index.
Identifiers: LCCN 2016039242 | ISBN 9781107152748
Subjects: LCSH: Astrogeology – Popular works. | Planetary science – Popular works. | Solar system – Popular works.
Classification: LCC QB454.B87 2017 | DDC 559.9 – dc23
LC record available at https://lccn.loc.gov/2016039242

ISBN 978-1-107-15274-8 Hardback

This book is dedicated to my parents
Ralph Julius Buratti (1920–2008)
Hildegarde Mahala Singles Buratti (b. 1922)

# Contents

*Color plates section can be found between pages* 118 *and* 119.

# Acknowledgments

I owe the most gratitude to those who have taught me important life lessons. My parents Ralph and Hildegarde Buratti were people who set an example by doing rather than by saying. My graduate school advisor, Professor Joseph Veverka, taught me that if it's worth doing, it's worth doing well. If it's not worth doing, don't even get started. I am grateful for the scientific colleagues and other friends who have stood by me throughout the years. Special mention goes to my fellow students at Cornell, Julio Magalhaes, Steve Lee, Jay Goguen, the late William Reid Thompson, the late Damon Simonelli, and Peter Thomas. We share a bond like siblinghood. My other best friends, Lisa Nuchtern, Ruth Several, Marilyn Andrews, and Sally Swigart, and my real siblings, Bruce Buratti and Brenda Buratti, were always there no matter what.

My editors at Cambridge University Press, Vince Higgs, Philippa Cole, and Lucy Edwards, provided just the right amount of prodding when other obligations became overwhelming. I was fortunate to have generous colleagues who reviewed chapters of the book: Don Yeomans, Rosaly Lopes, Anne Verbsicer, Jason Hofgartner, and Linda Morabito Kelly. The remaining errors are all mine of course. For interviews I thank Linda and Anne, as well as Tim Parker and Kenneth Lawrence at NASA's Jet Propulsion Laboratory (JPL).

The love and support of my family has been enduring and unconditional. My three wonderful boys (men now), Nathan, Reuben, and Aaron Buratti Lam, all accomplished in their own right, took the time to read early manuscripts and to offer suggestions. Aaron helped in the production of the manuscript at the very end. But foremost is my husband, Kai S. Lam, who set the example with his own "butt in chair" philosophy of writing, and continually encouraged me to "just write." He also provided the scientific honesty and measured perspective to tone down passages that sometimes veered off.

And of course I am indebted to NASA and JPL, where unbridled curiosity and superb leadership are manifest in the missions and research in which I have been so privileged to participate.

# Introduction

Writings of the great thinkers abound with words expressing the great hold of astronomy. Plato said "astronomy compels the soul to look upwards and leads us from this world to another." When William Herschel (1738–1822) – the father of modern observational astronomy – received the Royal Society's Copley Medal in 1781, the Society President and naturalist Joseph Banks, stated that "the treasures of heaven are well-known to be inexhaustible." Astronomers themselves have spoken of their unquenchable curiosity and their drive and persistence to slake that curiosity. German astronomer Johann Schroeter (1745–1816) spoke of the "impulse to observe," while another astronomer said the purpose of existence is to observe. When Herschel was once asked why he had become an astronomer (with the implication – familiar even today – that a life of observing is impractical, even useless) he simply said that when he looked up and saw the beauty and wonder of the skies he didn't understand why everyone wasn't an astronomer.

But then there is the counterpoint in the public mind, captured in Walt Whitman's (1819–1892) poem "When I Heard the Learned Astronomer":

> When I heard the learn'd astronomer,
> When the proofs, the figures, were ranged in columns before me,
> When I was shown the charts and the diagrams, to add, divide, and measure them,
> When I, sitting, heard the astronomer, where he lectured with much applause in the lecture-room,
> How soon, unaccountable, I became tired and sick,
> Till rising and gliding out, I wander'd off by myself,

In the mystical moist night-air, and from time to time,
Look'd up in perfect silence at the stars.

The calculating astronomer misses the essence of the thing. As brilliant as he was, Whitman was wrong on this one. Most non-scientists think science is dry, fact-based, memorization – exact, or impenetrable. It is none of those things. Science is an endeavor of creative thought and activity, and it affects our everyday world by paving the way for technological inventions and by providing the groundwork for everything from weather forecasting to curing cancer. In its highest form it is no different from poetry. To see for the first time an aspect of nature is the same as crafting a group of words that speak a deep truth that every sensitive person can relate to, but perhaps can't put into words. The ancient Greeks accepted the profound status of Astronomy by deeming it one of the four quadrivia, areas of knowledge that form the basis of wisdom (the others are Arithmetic, Geometry, and Music). Astronomy may be the most empirical of these four subjects, but the driving force behind it is part of the abstract invisible world that is the wellspring of human activity, the same drive that gives rise to great literature, music, art, and some might even argue the spiritual impulse.

How often I have heard from non-scientists, whenever I speculate a bit on anything: "You're a scientist, you know you have to be sure of everything and have absolute proof." Science isn't like that – it's based on hunches and what isn't immediately evident in the data before you. It is propelled by speculation, leaps of faith, doubt, and disagreement. And even when we think we are sure, paradigms come tumbling down. In graduate school I learned life arose in shallow seas on the early Earth: molecules were zapped by lightning and sunlight, which in turn formed amino acids from which life somehow arose. Nobel chemist Harold Urey and his graduate student Stanley Miller had performed a series of key experiments in the 1950s that formed the foundation for this idea. A few short years after I completed graduate school, that paradigm had been completely turned around: the consensus was that life arose instead in thermal vents in oceanic ridges (so-called "smokers") because the bacteria there had a primitive genetic code that seemed to be imprinted on all life. I was talking with my friend and colleague Penny Boston of New Mexico Tech on how this paradigm had shifted so drastically within my own lifetime and she said, "Oh, that's all wrong. I'm pretty sure that life arose in deep caves." In one of the hottest areas of scientific research, an area that spans biology, planetary science, and astronomy – the origin of life – we are really no closer to answers now than we were a half century ago. Without speculation, creativity, and, yes, sometimes what seem to be crazy ideas, we'll get nowhere. (At least we moved beyond spontaneous generation as a means to create life – and all because Louis Pasteur had done an experiment to disprove it.) Science is everchanging: science is based on what you know at the moment. If more evidence comes in, you have to change your view. But any idea in science must ultimately hold up to experimental data and verification. Otherwise, it is just crackpotism.

The essence of scientific discovery was succinctly captured in an essay in the December 22–29, 2014 *New Yorker* by physician Jerome Groopman, "Science operates around a core of uncertainty, within which lie setbacks, but also hope."

It is acceptable among the educated public (and that's most of us, now that the majority of Americans graduate from high school and attend college) to lack scientific literacy and – most definitely – math literacy. There are people who would be embarrassed to admit they hadn't read *Moby Dick* or at least a couple of recent literary bestsellers. But they will be dismissive about science, as if it is a world apart, definitely okay to know nothing about. Imagine my shock when I gave a little talk on the Moon in one of my kid's kindergarten classes and after the teacher (a graduate of the University of California, Berkeley) introduced me she added the insult "and she's good in Maaaath!" with a horribly wrinkled face. After a second of disbelief, I jumped up and asked the five-year olds how often you would have to cut an apple in half to have nothing left. They grasped the concept of infinity pretty easily. I told them mathematicians did cool and fun stuff like think about infinity. It seemed they found math ideas pretty easy and amazing. But I had these kids for 30 minutes and she had them for a full year. Shortly after this eye-opening experience, I wrote a proposal to NASA to do a Teachers' Workshop. I proposed a hands-on, inquiry-based workshop in astronomy to instill the flavor and meaning of science into teachers' curricula. The goal was to turn teachers and their students into mini-investigators. Many other scientists are doing similar things to communicate the wonder and joy of science, and the enterprise of discovery. I was reminded of the words of Dan Goldin, the Administrator of NASA at that time: kids are naturally interested in three things: dinosaurs, ghosts, and space, and we need to exploit that last interest.

Science teaches both critical thinking and quantitative thinking, and you cannot succeed in it – or understand it – unless you are curious and relentless in whatever you are doing. You will not discover anything if you give up. Those lessons apply to life as well. Most of us realize that excelling in sports requires teamwork, persistence, and hard work. Science requires all those things. Discovery can put you right into the "zone" of transcendence just as rowing, skating, running, wrestling, or playing baseball can. And as in sports – or music performance, or art, or writing – the pleasures of discovery enter dressed in the dour garments of drudgery, but they leave unexpected and sublime.

I hope that with this book I can impart some of the flavor of scientific discovery within my own field of planetary sciences. The exploration of our Solar System – and solar systems beyond – is a good launching pad for discovery because it is so interdisciplinary. It touches on the areas of physics, astronomy, geology, and even a little chemistry and biology. And then there are all the exciting – magical, sometimes it seems – areas of engineering that lie at the core of space exploration: rocket science, in other words. Planetary science covers the origins of planets and life, the structure and evolution of planets, moons, and all the small stuff like comets, asteroids, and dust that is out there. It's the story of where we came from and where we will end up. At some basic level, science is a collection of facts: the number of moons around the planets, the distances to the stars, that kind of thing. But our knowledge of those facts is ever-changing; the story of how we find out those facts is what is important.

In the past, astronomers studied individual objects – specific planets, moons, stars, and galaxies. More recently, as we've seen all the planets close-up, we now realize that the same physical processes occur on all the worlds explored. Now that thousands of planets have been discovered around stars other than our Sun, it makes more sense to look at planets as groups of objects: gaseous, rocky, icy, geologically active or inactive.

Instead of cataloguing a little bit about every planet and moon in the Solar System, like we are part of a cosmic coin collection, I've decided to talk just about the ones that I've worked on or that I find most interesting. And every time, I try to bring it home. I start in the inner Solar System and work outward. One theme in this book is to compare what we see on the Earth – the familiar – with what exists on the planets and their moons – the fantastic. This book isn't comprehensive in its coverage: I don't cover the gassy planets, dust, or magnetic fields in any detail. At some level the cosmos is indeed a coin collection, each piece comprising a unique part of the whole. Ignoring one item of this great collection is akin to the damage done by the extinction of a single species, or the disappearance of a single human language and the culture embodied within it. My aim is rather to pick a few moons and planets that are representative of the whole without diminishing it in any way.

This book is written for a layperson: there are no "prerequisites" beyond high school science.

With our discovery of thousands, and eventually millions, of solar systems around other stars, we now realize we are just one of many, "billions," of planets as my teacher and mentor Carl Sagan would say. To "seek out new worlds" is no longer a trope limited to science fiction. It is reality.

# 1 Mercury: The Hottest Little Place

To the casual skywatcher Mercury appears near the horizon just after sunset as a faint orange star bathed in the fading glow of the western sky. To the more dedicated observer, the planet also appears right before sunrise in the eastern sky. The ancients had two names for its dawn and dusk appearances: Apollo in the morning, to signify the appearance of the Sun, and Hermes in the evening, to acknowledge the Greek messenger god. The speed of Mercury's motion in its orbit – and as seen from the Earth – is faster than the other five planets easily visible to the naked eye. By the fourth century BCE, during the golden age of Greek experimental science, astronomers noticed that this faint planet appeared in the same position relative to the Sun at both dawn and dusk, and they realized the two apparitions were the same body. The Romans named the planet Mercury after their own swift messenger god. In Nordic mythology, Mercury was associated with Odin, or Wodin, from which Wednesday (Mercredi in French with similar renditions in the Romance languages) is derived.

Many astronomers have never seen Mercury, and the first sighting of this elusive, "mercurial" planet is always memorable. I still remember the night over a half-century ago when I stood alone in the middle of a corn field near my parents' house in Bethlehem, Pennsylvania and compared the great night sky to a tiny map I had cut out of the *Bethlehem Globe Times*. The Sun had shed its last ray, and I felt so small as I stood where the soft cusp of the field gave way to the harsh vastness of space. But I was reassured when I saw the little planet, blinking on and off, unmistakably where it should be.

Little experimental triumphs such as this one, when the smallness of our world and our concerns are dwarfed by the immensity and predictability of the stars and planets, were what drew me to the

FIGURE 1.1 Jupiter and the crescent Moon are the bright objects in the sky, with Mercury just visible above the haze along the horizon (to the right of the leftmost small tree). Image by Steve Edberg. See plate section for color version, where Mercury is more prominent.

study of the cosmos. I didn't see the planet again with my own eyes until the mid-1990, when I was a fully-fledged astronomer observing on the 200-inch Hale telescope at Palomar Mountain. My colleague Phil Nicholson of Cornell University and I went out onto the catwalk circling the dome to inspect the weather and observing conditions. Phil quietly pointed out that Mercury was visible, its disk bobbing in the thick atmosphere above the faintly lit western horizon. The deep silence and the canopy of stars surrounding the Californian mountain drew me back to that night when, as a child, I had stood on the edge of the cosmic shore to glimpse Mercury for the first time. Mercury, Jupiter, and the Moon appear in Figure 1.1, a picture taken just after sunset by my friend and colleague Steve Edberg, an engineer and an ace amateur astronomer at NASA's Jet Propulsion Laboratory.

Astronomers like dark, moonless skies where even the faintest objects step out of the abyss: Mercury's location so close to the blinding Sun means it is exceedingly difficult to study. Even after the Sun sets, Mercury lies close to the western horizon, the brightest part of the sky. In the tropical zones it's a little farther above the horizon, because the path of the planets goes closer to the top of the sky – the zenith, in astronomical terms. When it gets really dark, which astronomers call "astronomical twilight," typically about an hour after sunset (for the technically minded: this deep twilight occurs when the Sun is 15 degrees below the horizon), Mercury has already set, or it is peeking through hopelessly opaque haze on the horizon. When Mercury is visible in the east just before dawn, the planet rises in what is then the brightest part of the sky, just before the Sun makes its appearance.

So Mercury's perpetual location in the brightest part of twilight's firmament meant that not much was known about the planet until it was scrutinized by *Mariner 10* in 1974 and 1975 during three close flybys. But somehow Mercury has often found itself at the center of scientific advancement and controversy by serving as a kind of celestial experimental apparatus. The planet helped close the door on the European acceptance of the geocentric model of the Solar System, in which the planets and the Sun all orbit the Earth. In 1610 Galileo disproved this incorrect theory by carefully noting that Venus undergoes a full cycle of phases, waxing and waning from new "moon" to full moon and back (see Chapter 2). The planet could only exhibit this phenomenon if it orbited the Sun on a path inside the Earth's orbit. Galileo's telescope wasn't powerful enough to observe the phases of Mercury, the other planet that is interior to the Earth's orbit and thus goes through a full range of phases; and of course the great astronomer was stuck with the same dreadful observing conditions faced by astronomers today. But only 29 years later, still during Galileo's lifetime, Italian astronomer Giovanni Zupi (c. 1590–1650) used his slightly more powerful telescope to observe the phases of Mercury. This observation demonstrated conclusively that Mercury as well as Venus orbited the Sun and not the Earth.

Mercury also provided the first experimental clue to another great idea: Einstein's General Theory of Relativity. In the eighteenth century, scientists used Newton's laws of motion to calculate the

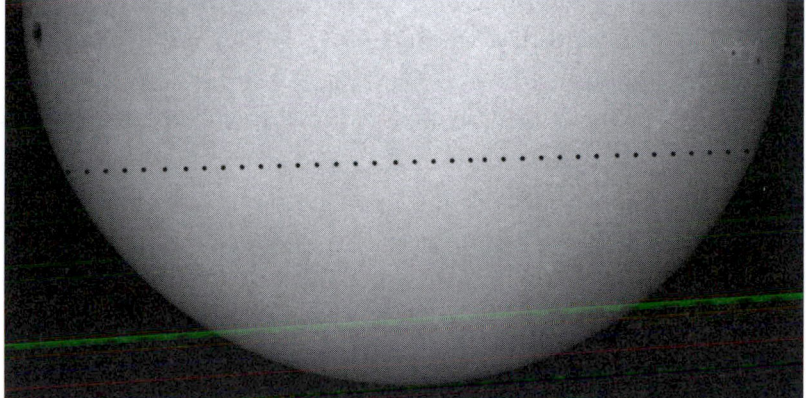

FIGURE 1.2 A transit of Mercury captured by NASA's Solar and Heliospheric Observatory (SOHO) on November 8, 2006. The entire event was about five hours long. Courtesy NASA.

times at which Mercury would pass exactly between the Earth and the Sun to appear as a dark spot moving across the face of the Sun. These so-called "transits," an example of which is shown in Figure 1.2, don't occur every time Mercury passes between the Earth and the Sun, mainly because the orbit of Mercury is inclined to the Earth's orbit. Earth and Mercury need to be at one of the two points at which their orbits cross – the "nodes" in technical jargon – at the same time for a transit to happen. To make things even more complicated, Mercury's orbit is elliptical – its distance from the Sun varies from about 29 to 43 million miles. Because the planet travels faster in its orbit when it is closer to the Sun, it was very challenging to calculate exactly when the transits of Mercury occurred. But the Golden Age of celestial mechanics was the nineteenth century: famous mathematicians took pride in their knowledge of Newton's laws of motion by predicting where the planets and moons and the Solar System's small bodies, such as comets and asteroids, would be at all times. And they did all these calculations without computers!

But the calculations for the times of the transits of Mercury were off by as much as an hour, even when the gravitational effects of all the known planets were taken into account. Mercury was known to exhibit a perplexing effect known as the advance of its peri-helion (the closest point in its orbit to the Sun): a slow rotation of its elliptical (egg-shaped) orbit in the direction of the planet's motion by about 0.16 degrees each century. Most of this advance could be explained by the gravitational pull on Mercury by the other planets, but a small amount remained unexplained. In 1843, Urbain Le Verrier (1811–1877), the French astronomer, mathematician, and co-discover of Neptune, calculated this unexplained amount to be about 38 arc seconds per century (the updated amount is 43 arc seconds, or 7.5% of the total; one degree is 3,600 arc seconds). It would take Mercury three million years for the orbit to advance to where it had started. A similar close analysis of the orbit of Uranus is what led Le Verrier to correctly predict the orbit and location of Neptune in 1846.

Why care about these transits and the times of their occurrence, beyond the somewhat dry analysis of bodies moving in space? Anyone who has ever witnessed a solar eclipse – a transit of the Moon in front of the Sun – and the period of anticipation that precedes the event, has experienced that tension that combines one's sense of smallness in the midst of a great cosmic occurrence with the feeling of triumph that we know exactly when it is coming. The transits of Mercury can only be seen through a telescope, so observing these events became a sort of astronomical status symbol in the eighteenth and nineteenth centuries. But there is a scientifically more substantial reason to watch these transits: they can tell us the size of the Solar System. The path of Mercury across the face of the Sun, as well as the times of the beginning and end of the transit, vary depending on one's location on the Earth. These variations depend on the distance to Mercury. If we know the distance to Mercury, we know the distances to all the planets. Kepler's third law says that the square of a planet's orbital period around the Sun divided by the cube of its mean distance from the Sun is a constant. The orbital periods of the planets around the Sun were well known, so if we knew the distance to just one planet, *all* the distances to each planet could be calculated. (If you don't think math is fun, powerful, significant, and useful, just ponder this point.)

It is easier to observe a transit of the larger, closer planet, Venus, so it was the prime planet used to measure the size of the Solar System, but because Mercury's orbital period around the Sun is shorter, its transits happen more often, and they too were closely observed. Transits are also useful for other scientific reasons: for example in 1769 astronomers noted that Mercury had little or no atmosphere, the only such planet known at that time. In a modern twist, on June 3, 2014, the *Curiosity* rover on Mars observed Mercury transiting the Sun; such transits of course occur at different times on Mars than on the Earth.

Trying to replicate his success at predicting the existence of Neptune, in 1859 Le Verrier explained the discrepancy in the advance of Mercury's orbit to an unknown planet closer to Sun, which was dubbed Vulcan, after the Roman god of fire, a fitting contrast to the god of the sea. (Mr. Spock's Vulcan is a planet orbiting around another star, 40 Eridani.) Why had Vulcan never been seen? Because it was even harder to see than Mercury, always lost in the blinding glare of the Sun. It would be possible to see easily only during a solar eclipse, when the bright light of the Sun is extinguished, or during a transit. No one had reported an unknown planet during an eclipse, but the dawn of astronomical photography did not occur until a century later, so there were no archived images of solar eclipses that could be inspected after the event. Transits presented an even greater problem: to predict them, one needed to know the orbit of a planet, and Le Verrier didn't have sufficient information to compute an orbit for Vulcan. Perhaps someone had seen something that was discounted, much as previous observations of Uranus had been made for a century prior to its discovery (it was not recognized as a planet in these previous sightings). Edmond Lescarbault (1814–1894), a French physician and amateur astronomer, reported that he had seen a dark object move across the Sun in 1845 that he was now certain was Vulcan. Based on this information, Le Verrier calculated a 20-day orbit for Vulcan and placed it at 13 million miles from the Sun. He went on to predict more transits, but none were ever unambiguously confirmed. This situation was at odds with the usual trajectory of scientific discovery: if something is right, it is affirmed by a growing avalanche of data. Astronomers broke into camps of those who denied Vulcan's existence, to those who upheld it, the latter led by Le Verrier, who remained a believer until he died in 1877.

The year after Le Verrier's death, the path of a solar eclipse was to pass over North America. On July 29, 1878, just about every American astronomer geared up his or her equipment to look for Vulcan (yes, there were women astronomers even then, including the famous Nantucketer, Maria Mitchell (1818–1889), a professor of astronomy at Vassar College at the time). The path of the eclipse's totality extended from Wyoming – which did not become a state until 1890 – to Texas, with a partial eclipse engulfing the entire country. Only two astronomers of any note saw something: James Craig Watson (1838–1880), who was the director of the Detroit Observatory, and Lewis Swift, director of his own observatory, reported sightings that seemed to be Vulcan. But this Vulcan was only 400 miles in diameter, much smaller than the Moon and not nearly large enough to cause the changes in Mercury's orbit. When astronomical photography stepped into the forefront as a research tool, other American astronomers wiped away any doubt regarding the veracity of Vulcan. In 1909, William Wallace Campbell (1862–1938), the Director of Lick Observatory, showed that there was nothing inside the orbit of Mercury larger than 30 miles in diameter, only one millionth the size required to explain Mercury's advancing orbit.[1]

Now it was time for Einstein to come to the rescue. He published his General Theory of Relativity in 1915. This theory predicted that the force of gravity – a property of all matter – bent space and time around it. The massive gravitational field of the Sun, with 99.9% of the mass of the Solar System, warped space around it such that "the orbital ellipse of a planet undergoes a slow rotation, in the direction of motion, of amount (equation follows) per revolution . . . Calculation gives for the planet Mercury a rotation of the orbit of 43 arc seconds per century." This prediction was exactly that required.[2]

We see this relativistic effect easily only at Mercury because of its proximity to the Sun and its short year of only 88 days. For example, the perihelion shift of Earth's orbit due to general relativity is 3.84 seconds of arc per century, and Venus's is 8.62. Another prediction of Einstein's theory is the bending of starlight by the Sun during an eclipse, which was observed by the British astronomer Arthur Eddington (1882–1944) in 1919. Einstein's third prediction of general relativity, the slowing of clocks in a gravitational field, was conclusively observed on the Earth in 1959 by Harvard Professor Robert Pound and his graduate student Glen Rebka. They placed two atomic clocklike devices in Harvard's Jefferson Laboratory, one in the basement and another in a tower six stories high. The device in the basement, where gravity was higher, ran more slowly than the one in the tower. Today corrections for this general relativistic effect are done routinely in everything from spacecraft orbital calculations to GPS coordinates.

The study of Mercury itself from the late eighteenth century until 1965 – nearly two centuries after the dawn of modern observational astronomy until the space age – illustrates one of the most confounding bugaboos of the scientific method: the bandwagon effect. Scientists are only human, and they impose their own prejudices and foregone conclusions on their experiments.

In 1800, Johann Schroeter (1745–1816), a German lawyer and astronomer, observed what he interpreted as 20-km high mountains on the surface of Mercury. Schroeter's motivation for observing the planets was to expand a notion that was popular at that time: the plurality of inhabited worlds. The Creator had made all the worlds in the cosmos for a purpose, and that purpose must be to have them teeming with inhabitants; otherwise there would be no reason for their creation. Schroeter also admitted to an "irresistible impulse to observe," and the historian of astronomy Agnes Clarke called him the "Herschel of Germany."[3] With the help of the self-taught mathematician and astronomer Friedrich Bessel (1784–1846), Schroeter determined an Earthlike rotation period for Mercury of 24 hours, although their determination of the tilt of the planet – 70 degrees – was decidedly

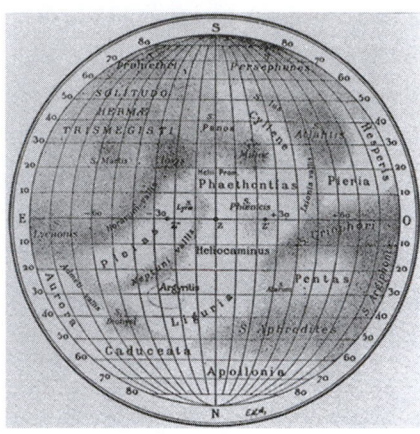

FIGURE 1.3 Antoniadi's map of Mercury, from which he observed an 88-day rotation period.

unlike Earth's tilt of 23 degrees. This view of Mercury stood for over eight decades: scientists just wanted to see Mercury as an "Earth above."

In the 1880s Giovanni Schiaparelli (1835–1910), who was famous for the mapping of "canali" on Mars (see Chapter 3), drew surface features that he thought reappeared every 88 days, the same as the orbital period of Mercury about the Sun. Thus Schiaparelli thought the planet rotated on its axis once for every orbit about the Sun, keeping the same face to the Sun, just as the Moon keeps the same face to the Earth. This state, known as synchronous rotation, is due to tidal forces acting on both liquids and solids to slow the rotation of Mercury about the Sun (or the Moon about the Earth). Synchronous rotation is the end state of this slowing effect. The mapping of Mercury was taken up by Eugene Antoniadi (1870–1944), another Mars enthusiast and master chess player, who continued to follow features on the Mercurian surface through its 88-day rotation period. He published a book in 1934 (*La Planete Mercure et la Rotation des Satellites*), which contained maps of the features (Figure 1.3). The eternal hell-like heat on the side facing the Sun and the corresponding frigid climate on the far side meant that Mercury was decidedly unEarthlike.

My own first encounter with Mercury as a world rather than as an orange dot in the sky occurred in 1965, when I read the science fiction short story by Issac Asimov, *Runaround*, in his famous collection "I, Robot," published in 1950 (*Runaround* originally appeared in the March, 1942 issue of *Astounding Science Fiction*). The book served as the inspiration for the movie of the same name starring Will Smith, although the movie plot doesn't follow any of the stories in Asimov's collection. *Runaround* proceeds under the assumption that Mercury is in synchronous rotation so that one side of the planet (the "sunside" in Asimov's story) was very hot and the other side was very cold. On the sunside existed some very unEarthlike features: pools of liquid selenium, which were mined by robots for use in electronic devices. In the late 1940s, the new burgeoning world of semiconductor devices used selenium; only later was this rare element supplanted by silicon. Asimov describes a hot foreboding world, strangely beautiful with its scenes of sparkling lakes, crunchy crystals underfoot, and rugged mountains casting sharp shadows of welcome shade. In 2015 – yes, the novel is describing the present time! – the bungling, but resourceful, engineering team of Powell and Donovan are sent to Mercury to resuscitate a robot-powered selenium mining operation on Mercury's sunside. Their latest robot model, nicknamed Speedy, soon falls into a defective operational mode of incessantly circling a pool of selenium rather than fetching the precious liquid. The team correctly figures out that Speedy is experiencing a conflict between two of Asimov's laws of robotics: #2, which says a robot must obey the orders of a human, and #3, which states a robot must do all it can to protect itself. Vapors that were corrosive to metallic robots' bodies were being outgassed by the selenium pool, so Speedy instinctively walked away from it in an attempt at self-preservation. Then Robotics law #2 kicked in, and he reapproached the pool to fetch the selenium as commanded by his human masters, only to move away to protect himself (sorry for assigning the male gender to the machine). This equilibrium was destroyed only when Powell went out onto the blazing surface of Mercury and convinced Speedy that he (Powell) was about to die. Rule #1, that a robot cannot cause a human being to come to harm, which trumped rules #2 and #3, kicked in and Speedy left his circle to rescue Powell. Asimov made a prescient point in this fun story: advancing technology invariably leads to design flaws. Speedy was built with a higher sense of self-preservation than previous less expensive machines, and this seemingly small change was what led to the operational bug. Asimov's story is also an early description of space mining, the exploitation of celestial bodies for natural resources.

I can't think of any plausible geologic scenario that would create selenium pools on Mercury. Selenium has a melting point below the surface temperature of Mercury, but selenium is rare, and it exists chemically bound to minerals. It would never exist amassed in one place. But Asimov does describe some accurate landscapes: "a towering cliff of a black basaltic rock" and "gray pumice, something like the Moon." In the 1940s and 1950s scientists didn't know whether the craters on the moon were volcanic or impact features. Volcanism was the favored mechanism for their formation, so Asimov described an igneous world. Even though now we believe the vast majority of craters not only on the Moon but in the Solar System are impact features formed by meteoroids and asteroids (see Chapter 4), the rocks on the surfaces of Mercury and the Moon are igneous. There are no liquids to help form sedimentary rocks, nor are there the terrestrial mountain-building processes that create metamorphic rocks. Asimov was correct: there is basalt on Mercury (and on the Moon as well). He also realized the heat of the surface would cause volatile materials to sublimate and create a tenuous atmosphere. "There is a thin exhalation that clings to its surface – vapors of the more volatile elements."

The belief that Mercury was in synchronous rotation stood for a period of eight decades. But as so often happens, established scientific fact was demolished by a new technological advance: radar. Radar was developed during World War II, primarily by the British to detect German bombers advancing toward the coast of Britain. Radar measures the distance and speed of a remote object, as close as a passing car and in principle as far as a distant galaxy. Soon after its development for military purposes, astronomers realized they could use radar to measure accurately the distances to planets and to determine how

FIGURE 1.4 Gordon Pettengill (left), now a retired MIT professor, who measured the velocity of Mercury in its orbit with the Arecibo telescope (right) and found that it didn't keep the same face toward the Sun. Courtesy of Gordon Pettengill and Arecibo Observatory, a National Science Foundation Facility. Arecibo photograph by David Parker, 1997/Science Photo Library.

fast they moved in their orbits and how fast they spun on their axes. The first radar signal was bounced off the surface of the Moon in 1957 by scientists at the US Naval Research Laboratory. Now scientists can measure the distance from Earth to the Moon – 238,000 miles on average – to about a tenth of an inch: this accuracy is the same as measuring the distance between New York City and Los Angeles to the width of a hair. In 1965, the American radar astronomers Gordon Pettengill and Rolf Dyce bounced a radar signal off the surface of Mercury. They sent the signal from the Arecibo radio dish in Puerto Rico (see Figure 1.4), which is built into a massive natural concavity, and discovered that its rotational speed implied it spun on its axis every 59 days, which meant it rotated three times for every two orbits. Mercury did not in fact keep the same face towards the Sun!

So why did Schiaparelli and Antoniadi – both careful observers – goof up so badly? Observing features on Mercury is especially tricky, even more difficult than viewing it with the unaided eye, as every two orbits (every 176 Earth days) around the Sun, the planet returns to the same position and the same face does point to the Sun.

To make matters worse, Mercury orbits about four times around the Sun each Earth year. Thus, after that second orbit is completed, and the feature being observed is again in the same place, Mercury is also again at elongation and easily observable from Earth. At other times the feature supposedly eternally pointing to the Sun is either in dark or on the side of Mercury not visible from Earth. When one factors in the cloudy nights, it becomes quite plausible that Antoniadi and Schiaparelli would have both observed a feature only when it was pointing towards the Sun. If either astronomer noticed a feature out of place – and this would happen if they were observing faithfully year after year – they would just discount it. The bandwagon effect is strong, as we have seen already in the eagerness of the astronomical community to accept the existence of Vulcan.

All experimental measurements have bad or anomalous data: data gets read or copied down incorrectly; a measurement is a statistical fluke (5 heads in a row in a coin toss occurs every 32 times on average); there is some effect the experimenter is not aware of (a plane or bird flies in front of the telescope; the detector heats up or has a flaw; equipment gets kicked; the list goes on). The scientific method is rooted on the interplay between theory and experiment: experiment spawns a theory or model to explain the data, and further predictions of the model and the collection of subsequent data refine the theory. Science is propelled when the experimentalist has a hunch that a specific theory is driving the data. It is a scientific skill to see through the bad data, to see the forest for the trees. Non-scientists think that scientists are rational, exacting, "calculating," and uncreative. Nothing could be further from the truth. In an essay written in 1959, Isaac Asimov (who was a professor of biochemistry at Boston University as well as a science fiction author) said the crux of scientific creativity is "not only people with a good background in a particular field, but also people capable of making a connection between item 1 and item 2 which might not ordinarily seem connected."[4] Science is all about hunches, looking at things in new ways, and bringing order to chaos. Schaparielli and Antoniadi observed the same feature pointing to the Sun to support their "hunch" of synchronous rotation and, after that, everything they saw supported it. Other scientists jumped on the bandwagon, which was easy because not too many people were observing Mercury. The bandwagon effect blinded the scientists to alternative theories.

Recent studies of Gregor Mendel's work on the traits of garden peas, which established the laws of genetic inheritance, show that he had culled data as well. Robert A. Millikan (1868–1953) who did the Nobel Prize-winning oil drop experiment computing the charge of an electron also abandoned data. (I have done this tricky experiment in a physics lab while an undergraduate student at MIT, and it is impossible to successfully do it without abandoning most data.) Some might even say Mendel and Millikan fudged data. In an even more striking illustration of the bandwagon effect, Millikan's value for the electron's charge was slightly in error – he had used a wrong value for the viscosity of air. But future experimenters all seemed to get Millikan's number. Having done the experiment myself I can see that they just picked those values that agreed with previous results. Schiaparelli's and Antoniadi's sense that Mercury was in synchronous rotation was a hunch, and it seemed to fit most of the data. The only difference between their and Mendel's and Millikan's experiments was that the model of synchronous rotation for Mercury was wrong.

Even with this new dynamical state for Mercury, the most extreme excursion of hot and cold for any planet occurs on its surface: at the poles, the temperature drops below –300 °F (–184 °C), while at the subsolar point (where the Sun is right overhead) it reaches 800 °F (426 °C). Mercury has only a trace of an atmosphere, so there is no blanket to buffer these extremes. Mercury is also unusually dense, about 5.4 grams per cubic centimeter (3.1 oz/cubic inch), comparable to Earth's 5.5 (3.2). The planet formed in a hot part of the Solar System, so it was able to accrete so-called refractory materials, such as iron and nickel, and they tend to be dense. Most of the iron is in the interior of the planet, forming a core that comprises about 40% of its volume. (For comparison Earth's core is about 16% of its volume.)

FIGURE 1.5 A global view of Mercury, constructed by images obtained by the *Mariner 10* spacecraft about six hours before closest approach. Mercury has a diameter of 3,026 miles. The image was put together by Joel Mosher of NASA's Jet Propulsion Laboratory, Caltech.

Like the Earth, Mercury melted early in its history to differentiate into a core made of dense materials, a mantle, and a crust of lighter rocks.

The first close-up of Mercury's surface was obtained in 1974, when *Mariner 10* sailed past the planet three times (Figure 1.5). The *Mariner* program was already mature by then, so the exploration of Mercury didn't suffer the same series of failures that plagued the exploration of Mars and Venus (see Chapters 2 and 3). The mosaic constructed of images from the spacecraft's cameras shows a decidedly lunar-like planet. (I like to fool people when I give talks and ask them to identify Mercury. Except for the amateur astronomers who populate Astronomy Clubs, who know more facts about planets and stars than I do, almost everyone identifies Mercury as the Moon.) Modern planetary scientists weren't disappointed when they saw the cratered face of Mercury as they were when the craters on Mars were first revealed by *Mariner 4* in 1965. They had never pictured Mercury as an Earthlike world that was an abode for life. *Mariner 10* discovered Mercury's tenuous atmosphere, presaged by Asimov, and a weak magnetic field. Scientists now think magnetic fields on the "terrestrial" planets (Mercury, Venus, Earth, and Mars) are caused by a combination of a liquid iron core and planetary rotation setting up a dynamo. Earth and Mercury are the only two of the four above planets that have these two attributes, although Mercury rotates only once every 59 days on its axis; perhaps this slowness is responsible for its magnetic field being only 1% as strong as Earth's. But the field is strong enough to hold the "solar wind" at bay. This supersonic outflow of charged particles from the Sun engulfs the Solar System's entire family of planets. Unless a planet possesses a magnetic field or atmosphere to deflect these dangerous particles, they impact the surface to render it inhospitable for life (among other things – just notice how the Sun's activity affects radio reception). *Mariner 10* showed that Mercury has a large impact structure, which was named the Caloris Basin: this 1,000-mile wide feature was caused by the collision of a large body (about 60 miles or more) nearly 4 billion years ago. Mercury has no moons; Venus is the only other planet that lacks companions, but Kepler predicted it should have one for "esthetic" reasons.

There are subtle differences between the Moon and Mercury. They both have two major types of terrain: cratered and smooth or plain-like, but Mercury's plains are hummocky or crinkled, implying that the planet may have slightly decreased in size in the past, such as when it cooled (see Figure 1.6). Mercury is denser than the Moon, so it must be made of heavier materials. Mercury has a larger iron core than the Moon, which is probably why it has a magnetic field while the Moon has only a trace of one.

But the similarity between the surfaces of Mercury and the Moon tells a significant story about the early history of the Solar System. Scientists long realized that without erosional processes to erase the face of the Moon as they do on Earth, the early history of the Solar System is written on the Moon's surface. But scientists were not

FIGURE 1.6 A *MESSENGER* image of Mercury showing hummocky terrain that suggests the planet contracted early in its history, perhaps from cooling. The large fold ("escarpment" in geologic terms) in the middle of the image is Beagle Rupes, named after Charles Darwin's ship. The image is about 490 miles across. NASA/Johns Hopkins University Applied Physics Laboratory/Carnegie Institute of Washington.

sure if that history might only apply to the Moon. When scientists glimpsed the cratered face of Mercury, they saw that both bodies spoke of the early history of the inner Solar System. Later probes extended this history to the entire Solar System. It was a violent history, one of large impacts and subsequent melting. The Solar System accreted out of a cloud of gas and dust. At some point, small rocky planetesimals formed out of the chaos, and they began to accrete together to form planets. But many of these planetesimals still remained after the planets came together. For the first few hundred million years of the Solar System's life, they impacted the planets and recorded their presence in the form of impact basins and craters. *Mariner 10* showed that this picture of the early Solar System extended beyond the Moon. This early violent period, which planetary scientists call the Period of Heavy Bombardment, extended from the formation of the planets 4.65 billion years ago to about 3.8 billion years ago. The collisions petered out during the tail end of the Late Heavy Bombardment, but they never completely ceased, as we shall see in Chapter 4. But, as the second battered surface in the inner Solar System, Mercury again formed our view of the "big picture," posing as a great laboratory in the sky.

A world where lead would melt: barren and almost airless, encircled by just a scrim of vapor, with little erosion . . . not very Earthlike

FIGURE 1.7 A *MESSENGER* mosaic of Mercury. The bright patch in the upper-right section of the planet is the Caloris Basin. NASA/Johns Hopkins University Applied Physics Laboratory/Carnegie Inst. Washington.

at all. So scientists and the public alike were amazed when, in 1992, Marty Slade, Bryan Butler, and Dewey Muhleman of Caltech and JPL used the Goldstone radio telescope, part of NASA's array of telescopes that comprise the Deep Space Network for tracking distant spacecraft, to detect ice near the poles of Mercury. Many scientists were at first skeptical, saying that perhaps it wasn't ice the radar astronomers were seeing but some other substance such as sulfur that is highly reflective at radar wavelengths. The observation has stood all assaults, and more recently ice has been detected at the poles of the Moon, where temperatures drop even lower, particularly in permanently shadowed regions of crater floors. Where does the ice on Mercury (and the Moon) come from? It has either outgassed from their interiors, or it is transported to the surface by impacts from ice-rich comets. Most likely, both sources are responsible for the ice on these barren worlds.

Mercury revealed more of itself during its exploration by the *MESSENGER* mission (MErcury Surface, Space ENvironment, GEochemistry, and Ranging, showing that NASA's acronyms are getting ever more clever), one of NASA's line of lower cost *Discovery* missions. Like *Mariner 10*, *MESSENGER* flew by the planet three times, but then it went into orbit for a detailed investigation, imaging 100% of the surface; *Mariner 10* captured only 41% of the planet. A planet-wide mosaic of Mercury is shown in Figure 1.7. Higher

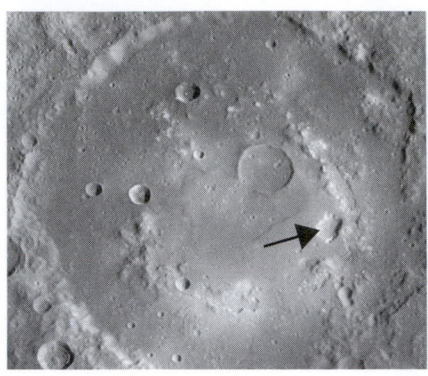

FIGURE 1.8 A *MESSENGER* image of the Praxiteles crater on Mercury, one region identified by Dave Blewett and his colleagues as a possible outgassing site. The hollow identified by the arrow is a possible vent surrounded by a bright deposit that may be volcanic ash. The crater is about 113 miles in diameter. NASA/ Johns Hopkins University Applied Physics Laboratory/Carnegie Institute of Washington. See plate section for color version.

resolution images of what geologists call lobate scarps provided evidence that Mercury shrank even more than shown by *Mariner 10* (see Figure 1.5).

One question that intrigues scientists and the public alike is whether Mercury might still be active. Mercury does look as dead as a planet can be. But some interesting evidence has surfaced that even if full-scale volcanoes are not erupting on the surface, little puffs of gas might still be coming from its interior. Dave Blewett and his colleagues at Applied Physics Laboratory noticed that some of *MESSENGER'S* high-resolution images of Mercury's impact craters revealed low-lying hollows surrounded by bright deposits (Figure 1.8). Blewett et al. hypothesize that these hollows are locations of outgassing, and the bright deposits are products of that outgassing, perhaps even ash deposits. Presumably, the outgassing occurs only in the craters because that is where the crust is broken and weak. This discovery of possible activity on Mercury is not just a curiosity. If the planet is still outgassing after more than 4 billion years, it must have harbored many more volatiles than previously thought. The formation of the Solar System was not just a simple picture of the refractory materials such as nickel and iron condensing to form the inner planets, with all the volatile material condensing in the outer regions. Again, Mercury is trying to tell us something fundamental about the world, namely how the Solar System formed.

NOTES

1 Asimov, I. 2014. "Isaac Asimov asks, 'How do people get new ideas?'" *Technology Review* Oct. 24, 2014. (www.technologyreview.com/view/531911/isaac-asimov-mulls-how-do-people-get-new-ideas).

2 Lorentz, H. A., Einstein, A., and Minkowski, H. 1916. "The Foundation of the General Theory of Relativity" *Annalen der Physik* 49, 769–822. A translated version is cited in *"The Principal of Relativity"* by J. Weyl (translated by W. Perrett and G. B. Jeffery) Dover New York 1952.

3 Sheehan, W. and R. Baum 1995. "Observation and inference: Johann Hieronymous Schroeter 1745–1816." *Journal of the British Astronomical Association* 105, 171–175.

4 Asimov, I. 1975 "The planet that wasn't." *The Magazine of Fantasy and Science Fiction*. May 1975.

# 2 Venus: An Even Hotter Place

I was in the third grade, at home sick with the measles in the era before the MMR vaccine. I was reading a book my parents had bought for me and my brother Bruce, *A Child's Book of Stars* by Sy Barlowe, and I became fascinated by a picture on the first page showing a caveman looking up at the sky. The fur-clad Neanderthal peered out into the wondrous, star-studded night sky, blacker than anything we have ever seen because all those shopping malls and street lights weren't to appear for tens of thousands of years. But then I turned a few pages, and I was drawn into a fantasy that changed my life. A veiled, crescent Venus loomed on one corner of the page, but at the bottom was a scene of what the surface of Venus might look like. A low fog clung to rugged canyon lands, and in the foreground were fern trees and exotic plants. "The temperature is possibly not too extremely hot nor too extremely cold and plants and animals could live there," the text declared (Figure 2.1). The landscape just pulled me in – I could imagine being there and exploring this tropical world. Enclosed in my little bedroom, I was simply awed by this enlargement of my world. The idea was firmly planted in my mind that one day we must go to this planet.

Many other children's books showed a lush Venus. *Space Cat Visits Venus* by Ruthven Todd has Space Cat and his human guide – a gun-toting cowboy type – cavorting among exotic jungle plants. *Encounter Near Venus* by Leonard Wibberley shows a flying saucer taking off from a swamp. Heinlein's *Podcayne of Mars* takes place partly on Venus, but this Venus is run by a latter-day monopoly where money isn't only the only currency, but the only everything. In adult science fiction, just about every human activity is replicated on Venus, in stories that underscore both the scientific and cultural

SOME SCIENTISTS TELL US THIS IS WHAT WE MIGHT FIND IF WE WERE TO HAVE A CLOSE-UP VIEW OF THE SURFACE OF CLOUD-SHROUDED VENUS

FIGURE 2.1 From *A Child's Book of the Stars* by Sy Barlowe (1953).

labels of their times and their authors. Under the pen name William Tenn, UFO researcher Philip Klass wrote the short story *On Venus Have We Got a Rabbi!* – which takes the "who is a Jew?" question into outer space. At the First Interstellar Neozionist Conference on a dusty arid Venus almost a millennium in the future, a group of aliens from a planet orbiting the star Rigel claim descent from Jewish colonists and want to be included as delegates to the conference (the story never explains how humans morphed into brown, pillow-like giant spotted slugs with tentacles). In the lenient and ingenious style of rabbinic decisions, the Venusian Rabbi Smallman, a member of the "Levittown faction" of human Jews on Venus, crafts a compromise to include these ETs.

Venus is the only planet named after a goddess, the goddess of love and beauty, and this eponym appears to be common to almost all cultures. As difficult as Mercury is to see in the twilight sky, the eye of Venus glares right back at you in the most palpable way as the Morning or Evening Star. Because of its proximity to the Sun, Venus appears only in the sky shortly after sunset or before dawn, just as Mercury does, except it rises higher above the horizon (Figure 2.2). Next to the Sun and the Moon – and the occasional bright comet or

FIGURE 2.2 Venus over the Pacific Ocean. Image courtesy of NASA. Photograph taken by Mina Zinkova. See plate section for color version.

meteor – it is the brightest object in the sky, casting enough light to make newspaper headlines readable and to form shadows.

Venus is the celestial body most often misidentified as a UFO: an incident in 1969 at a Lion's Club meeting in Atlanta involved a sighting by President Jimmy Carter, although Carter says he was sure the object he saw was not Venus. As our closest planet and the one most similar to us in size (Earth is only 5% larger than Venus), Venus earned the moniker of our Sister Planet, and as such it has been the receptacle of fancy as well as beauty. Immanuel Velikovsky (1895–1979), a Russian–American psychiatrist and peddler of pseudo-science, wrote in his 1950 book *Worlds in Collision* that Venus started out as a comet generated by Jupiter and ended up in a regular orbit around the Sun, after causing the parting of the Red Sea and Joshua's biblical Sun to stand still. There is, of course, no evidence for any of these ideas. Scientologists believe Venus is the location for erasing all human memories prior to reincarnation, similar to the Greek river of Lethe. Venus and its alien astronauts turn up as the center of attention in many UFO cults, including that of the original hoaxer

George Adamski, who had "Nordic aliens" from Venus ferrying him around the Solar System. When I was a teaching assistant for Carl Sagan's astronomy class at Cornell, he assigned a debunking project to each student. One of our students created photographs of UFOs more convincing than those of Adamski by taking pictures of the top of a garbage can tossed into the air.

The Babylonians were the first to realize Venus was both the Morning Star and the Evening Star. The Greeks didn't realize this simple fact of astronomy until Pythagoras in the sixth century BCE; they had different names for the morning and evening apparitions, Phosphorus and Hesperus, respectively. The Babylonian *Venus Tablet of Ammisaduqa* (17th century BCE) records in detail the rising and setting times of Venus for 21 years. The Babylonians called Venus the bright Queen of Heaven, which is replicated in the Bible as Meleket ha-Shamayim (the Queen of Heaven in Hebrew); in rabbinic texts the planet was known by the less idolatrous name Nogah, "she who glows," and in modern Hebrew the planet is called simply Venus. Besides Saturn, Venus is the only planet to be mentioned in the Bible, except the Earth of course.

Venus possessed a supreme position in the Mayan faith, in their view of the world, and in their calendar. Like the Babylonians, the Mayans realized the Morning and Evening Stars were the same object. By observing Venus over many years, Mayan astronomers calculated the period of 584 days that Venus returned to the same position as seen from the Earth to an accuracy of 0.014%: nothing by today's standards, but not bad for naked-eye astronomy. The Venus cycle was used to predict auspicious times for national events, including wars. Many Mayan buildings are aligned with the two elongations of Venus, when it rises at its most northern or southern position. The fixation on Venus in the Mayan world view lies in its seeming victory over the Sun: just as the Evening Star appears to be "gobbled up" by the Sun as the planet appears to move closer and closer to the Sun as it orbits around it, the Morning Star appears "victorious" in the morning as it escapes the Sun's grasp. In China, Venus is known as "Jin Xing," the metal star, named after one of the five elements in classical Chinese cosmology (earth, water, wood, fire, and metal).

The observation of the motions of Venus and Mercury with respect to the Sun laid the groundwork for the heliocentric model of the Solar System. Many of us learned in school that Copernicus was the first advocate of the heliocentric model. This idea is simply untrue: Copernicus restored the idea of the planets orbiting the Sun that was developed by the early Greek experimentalists and carried along by Muslim astronomers in Moorish Spain and elsewhere. Pythagoras (sixth century BCE) understood that the Earth is a sphere – it was the round shadow of the Earth cast on the Moon during a lunar eclipse that convinced him – and that day and night are due to the rotation of the Earth. Other early Greek empiricists constructed a realistic view of the world piece by piece based on evidence and observation. Anaxagoras (fifth century BCE) deduced the correct ideas for the cause of lunar phases, and by the third century BCE Aristarchus of Samos developed a fully heliocentric model in which all the six known planets, including the Earth, orbited the Sun. In the same century, Eratosthenes performed an experiment using simple geometry to measure the size of the Earth to within 50 miles. Christopher Columbus read Eratosthenes, but he did not believe the results; if he did he would have known that the West Indies were not Asia. According to the Irish classicist and historian of science Benjamin Farrington (1891–1974), the rise of slavery and its disdain of manual labor (e.g., experimentation) caused the consensus view to shift to the more philosophically appealing cosmic model of Ptolemy, which put the Earth at the center of the Universe.

Like Mercury, Venus played a role as a laboratory in the sky. Galileo's observation in 1610 of the phases of Venus was the killer observation that banged the largest nail in the coffin of the European geocentric view of the Solar System. For Venus to be half-lit when it was furthest in the sky from the Sun, and to be fully lit when it was behind the Sun, and a crescent when it was nearly in front could only happen if Venus orbited the Sun inside of the Earth (see Figure 2.3).

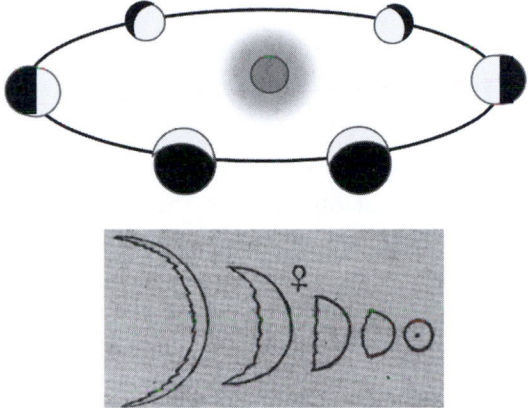

FIGURE 2.3 The top diagram shows the appearance of Venus as seen from the Earth in one orbit around the Sun. Like the Moon, Venus executes a full series of phases from fully lit to fully dark. This phenomenon could only happen if it orbited the Sun. The figure below shows Galileo's drawings of the phases of Venus. The crescent phases appear larger because they occur when Venus is closer.

Another dramatic example of Venus as a laboratory in the sky is that of the transits of Venus, which enabled the measurement of the distance between the Sun and the Earth, known as the astronomical unit (AU). As we saw in Chapter 1, once we know the distance between the Sun and one planet, we know the distance to all the planets, provided we know their orbital periods.

Transits of Venus are rarer than those of Mercury, simply because its orbital period is longer so that it has fewer opportunities to pass between the Sun and the Earth. But Venus is closer to Earth than Mercury, so its position on the face of the Sun at different places from the Earth varies more than that of Mercury. To see an example of how this concept works, put your finger six inches in front of your face, and view with one eye closed and then the other. Your finger is Venus and your two eyes are different telescopes on opposite sides of the Earth observing Venus during a transit. Notice how the background behind your finger changes. If you can measure this difference (or in the case of the real Venus, the difference in its position on the Sun from the two telescopes on Earth), you can also measure the distance to Venus, with a little high school trigonometry. Now move your finger 12 inches away to represent Mercury at a greater distance. When you view "Mercury" with each eye, the space it displaces behind your finger is less than that displaced by "Venus." Thus the position of Mercury on the Sun as seen by the two telescopes doesn't vary as much as that of Venus. In science, it's generally more difficult to measure very small numbers, so it was possible to measure the distance to Venus more accurately than the distance to Mercury.

Venus and Earth line up every eight years, but transits occur not even every 100 years. Why? The orbit of Venus is inclined about 3 degrees to that of Earth, so similar to the case of Mercury, not only do Venus and the Earth have to line up, they have to be where the plane of the planets' orbits cross, the nodes. This configuration happens on a 243-year cycle (when they are back at the same node) in pairs eight years apart every 105.5 and 121.5 years. Two occur at a time because the Sun isn't a point source: it covers enough of the node so that when Venus returns in 8 years to line up again, the Sun is still close enough to the node so that Venus "hits" it. The next transit will not occur until December 2117 (may you live that long in good health).

Although the Mayans and Babylonians observed Venus closely there is no record of a transit in their surviving writings. The first recorded observation of a transit is probably that of Ibn Sina (Avicenna), a Persian scientist and philosopher, who in 1032 CE referred to Venus as "a mole on the face of the Sun." Kepler predicted that a transit would occur in 1631, but his techniques weren't accurate enough to define where on Earth it would be visible: like solar eclipses, transits cast a path of visibility on the Earth, although the transits are visible over a much larger area because the cross-section of Venus – which determines the size of its shadow – is larger than that of the Moon. In 1639, the English astronomer Jeremiah Horrocks (1618–1641) became the first person to both predict and observe a transit of Venus. He calculated the distance from the Earth to the Sun to be about 60 million miles (the actual number is 93 million miles). The next pair of transits, in 1761 and 1769, represented one of the first occasions of a cooperative international scientific expedition. Astronomers traveled all over the world – possibly as the first example of a scientific travel junket – with telescopes and equipment in tow. The fledgling scientific community in the American colonies built three observatories for the event, and the first volume of the *Transactions of the American Philosophical Society* (founded by Benjamin Franklin and John Bartram), published in 1771, described the results. Just as the transit of Mercury in 1769 showed that planet lacked an atmosphere, the 1761 transit showed Venus had an atmosphere. The Russian astronomer Mikhail Lomonosov (1711–1765) noticed an arc of light extending from Venus as it exited the transit: he deduced that it was light from the Sun refracted through the atmosphere.

Perhaps the sorriest saga of the many journeys carried out to study this pair of transits was that of the French astronomer Guillaume Le Gentil (1725–1792), who sailed to Pondicherry, India, a French outpost. When he arrived he discovered the city was occupied by the British. He was still at sea when the date of the transit arrived, but he decided that since he had invested so much time and so many assets in the transit, he would stick around for the second one. He went back to Pondicherry, which had since been reoccupied by the French, and built an observatory. He was clouded out on the day of the transit, and returned home in defeat. Plagued by illness and storms, the return trip to Paris took two years, making his absence a total of 11 years. When he arrived home he learned that he had been declared dead, that his wife had remarried, and that his property had been possessed by his heirs. After all the drama, astronomers were able to calculate the distance between the Earth and the Sun to an accuracy of about 2%.

Although the distance to Venus can now be much more accurately measured with radar, the transits of 2004 and 2012 represented a second life for Venus as a laboratory in the sky to study planets in other solar systems. By the turn of the millennium, astronomers had discovered scores of planets around other stars, known as "exoplanets" or "extra-solar planets" (see Chapter 10), and they realized that when these planets transit in front of their own suns, they could be observed with very large sensitive telescopes on Earth. By carefully observing how starlight passes through the exoplanets' atmospheres, the physical character of these atmospheres – their pressure, composition, and whether they contain life-sustaining substances such as water and oxygen – could be studied. Since we already know quite a bit about Venus's atmosphere, its transits provide observations of a known object close up. These data represent a touchstone, a groundwork of verified data from a well-studied object by which to compare the results of exoplanets' transits. The main gas in the atmosphere of Venus, carbon dioxide, has already been detected on exoplanets, so a Venusian transit offers a realistic view for at least these exoplanets.

In the 1790s, Johann Schroeter tried to impose an Earthlike rotation period of 23 hours 21 minutes on Venus, similar to his findings for Mercury (see Chapter 1). Because Venus has a thick atmosphere that is nearly 100 times as dense as the Earth's (as stated above it is composed primarily of carbon dioxide rather than nitrogen and oxygen), its surface features are invisible. Schroeter could not possibly have tracked features moving on its surface as it revolved. He was trying to impose his view of a myriad of Earths above, all teeming with life, on whatever he saw. But it was hard to shake our view of Venus as Earth's sister planet. Other observers either confirmed his results or presented inconclusive counter-evidence.

The first devastating blow came when Gordon Pettengill and his colleagues beamed MIT's Millstone radar up to the planet in 1961 and found that the planet hardly rotated at all – it seemed to be in synchronous rotation, keeping the same face towards the Sun (this same observation placed an accuracy of 0.00001% on our knowledge of the astronomical unit). By 1967 the team had determined that Venus's rotational period was 245 days – more than its orbital period of 225 days. This observation means that Venus appears to slowly rotate backwards, a dynamical state that astronomers call retrograde motion. All the planets orbit around the Sun in a counter clockwise ("prograde") motion as seen from the Sun's North Pole. Except for Venus and Uranus, the planets also rotate on their axes in a prograde motion, with the Sun rising in the east and setting in the west. With its axis of rotation tilted 120 degrees, Pluto is also retrograde. Venus's strange, very unEarthlike state may be the compromise between two tidal phenomena that slow its rotation: the tidal force exerted by the Sun, and the tidal dissipation occurring in its thick, hot atmosphere.

Radio waves delivered the second of two punches – wrecking ball impact might be a better description for the second of the one-two punch – to our fantasy of Venus as Earthlike. In 1962, Carl Sagan published a paper in the newly minted journal *Icarus*, devoted to planetary science, that said radio-frequency observations of Venus could be explained if the surface temperature were a whopping 890°F (477°C). Venus was hotter than Mercury! Current best estimates place the temperature about 20 degrees lower. Furthermore, the high temperature was due to a runaway greenhouse effect caused by the trapping of solar radiation by the atmosphere's carbon dioxide, similar to the global warming effect that is occurring on the Earth today. Previous observers were unable to find even a trace of water on Venus, and now that observation made complete sense. Venus is so hot that all its water has evaporated and escaped out into space: a sobering thought when one considers that Venus may be a more evolved Earth.

Pettengill was already a legendary figure when I took his radar astronomy class at MIT in the mid-1970s. He was also a superb teacher: I could never understand the common criticism that research universities were packed with elite professors who cared little about teaching. Even as a naïve student, I could sense that professors who "punted" on lectures and student-time were looked down upon by their colleagues. But my most vivid memory of Pettengill's class was of fellow student Steve Ostro, who later became a leading radar astronomer in his own right (see Chapter 8). For the first test, which was an in-class open-book exam, Ostro walked into class carrying about a dozen books so that he would be prepared for anything.

He sat down and finished the test in about half the allotted time without ever opening any of the books. The rest of us struggled until the very end of class.

After a lag of a few years, science fiction started to depict Venus in a different light. It was a hell, where people lived in tunnels and bubbles. Larry Niven's *Becalmed in Hell* (1965) describes a hot Venus with a thick atmosphere and an "eternal searing black calm." The astronaut Howie and his robot companion Eric, who has a brain from a disembodied injured man that is connected to and controls the engineering systems of their craft, are unable to command their spaceship to move through the clouds of Venus. Howie blames the problem on Eric, who he believes has a psychological problem akin to PTSD that prevents him from operating the ship's rockets. However, when the crew returns to Earth, they discover that one of the ship's ramjets was jammed by the dense Venusian atmosphere. On the cover of *Venus of Shadows*, Pamela Sargent depicts a world of fire (although without oxygen there would of course be no fire), where people live in glass-enclosed cities. It is also notable for its depiction of a multicultural female, a far cry from the latter-day cowboy portrayed in Ruthven Todd's "Space Cat Visits Venus." In many stories, thoughts of terraforming Venus – engineering it to be more like the Earth – seemed to take the edge off of our disappointment. In *Becalmed in Hell*, Howie fantasizes turning off gravity so 99% of Venus's atmosphere escapes, ending greenhouse warming and restoring Venus to terrestrial temperatures. But he never mentions how to get the water back.

An armada of real spacecraft, both Soviet and American, stepped in and continued to peel away the layers of likeness between Venus and Earth. But the missions were plagued with problems. *Mariner 1*, our first interplanetary mission, went off course shortly after its launch on July 22, 1962 from Cape Canaveral and flight controllers commanded it to self-destruct. A little over a month later, *Mariner 2* was launched and flew by Venus on December 14, 1962 at a closest approach of about 22,000 miles. *Mariner 2* didn't carry a camera: its main goals were to study the magnetic fields and particles in the vicinity of Venus and to measure temperatures. *Mariner 2* confirmed the high temperatures of the planet's surface and showed that heat is transported easily from the sunlit side to the dark side. Venus had no magnetic field greater than 10% of Earth's. This result made sense in light of Venus's slow rotation.

Meanwhile, the Soviet Union was competing with the United States to build even more spectacular interplanetary missions. It launched *Sputnik 7* on February 4, 1961 for a crash landing on Venus, and *Venera 1* eight days later toward a flyby mission. Both spacecraft failed. These double disappointments were followed by a staggering 13, possibly 14, additional failures.[1] Finally, on March 1, 1966, *Venera 3* became the first spacecraft to crash land on a planet, and on October 18, 1967, *Venera 4* scored another first when it took data on the pressure and temperature of the planet's atmosphere. Originally, the Soviet Union had claimed *Venera 4* had soft landed on the surface, but when the US's *Mariner 5* arrived a day later and measured the surface pressure at 75 to 100 atmospheres, the Soviets had to retract their claim. Their craft was built to withstand pressures of 25 atmospheres at most. Carl Sagan's classic paper from 1962 already computed a surface pressure of about 50 atmospheres (the current estimate is 93 to 94 atmospheres); any engineer would design their spacecraft to withstand at least twice the expected conditions. *Venera 5* and *Venera 6* had already been built to the lower, incorrect tolerance, so they were designated as atmospheric probes instead of landers, sending back about an hour of data each in 1969. John Lewis, my brilliant and colorful professor for the course "Planetary Physics and Chemistry" at MIT in 1972, joked about the "gold bust of Lenin" encased in the *Veneras* slowly melting during their descent into the atmosphere. The Soviets executed a successful landing mission with *Venera 7* in 1970, and continued to send surface landers with ever increasing sophistication and success. In 1975, *Venera 9* sent back the first image of the surface of Venus. The Soviet landing missions culminated in 1981 with *Venera 14*, which analyzed a sample of basalt, a volcanic rock that

FIGURE 2.4 The surface of Venus as seen from *Venera 14*. Credit: Russian Academy of Science, posted on Reddit.

is common on the Earth, Moon, and Mercury, and sent back images, one of which is shown in Figure 2.4. The original color image shows volcanic rocks, and landscape that appears to be eroded by wind.

Landers investigate tiny little plots of land embedded on great worlds. Before scientists send landers to planets or moons, they typically reconnoiter the place by executing flybys and putting orbiters in place. Engineers find flybys are easier to design and fly than orbiters: there is no slowing down and setting off a complicated series of jet-firings to precisely counteract the gravitational pull of a body to place the spacecraft into the sweet spot of an orbit. But flybys take a snapshot in time, and often they observe less than half of the planet, as some of its surface may be in darkness (although that doesn't bother infrared instruments) or simply hidden. Orbiters not only case the whole planet, they can seek changes such as active volcanoes, disappearing polar caps, and clouds coming and going. And for a future explorer, they can map the entire body and select the best place for landing. Imagine if you had to understand the Earth based on data returned from an immobile robot that landed in the Gobi Desert, or in the midst of a tropical rainforest, or on Antarctica.

The problem with Venus is that its surface isn't visible because its thick atmosphere shuts out virtually all light, rendering flybys and orbiters useless for geological studies. Unless, of course, wavelengths other than visible light are used. In 1978, the United States stepped out in front of the Soviet Union by inserting into orbit around Venus a spacecraft known as *Pioneer Venus* for a 14-year mission. Again, radar came to the rescue. Among the 17 instruments on board was a radar mapper that could cut through the thick clouds and get a global view of Venus's surface. The mapper data was compiled into an altimetry map that showed two large plateaus, appropriately named Ishtar and Aphrodite, the first in the northern regions of the planet and the second near the equator. Located on Ishtar was an enormous mountain, Maxwell Montes, the only feature on Venus named after a man, James Clerk Maxwell, who developed the fundamental laws of electromagnetic theory. The mountain is almost 7 miles high, or more than a mile higher than Mount Everest. And it appears to have a sort of "frost" on its summit: like other high-lying regions of Venus, Maxwell Montes appears to possess a patina of a material that is highly reflective at radio wavelengths. Exotic materials such as pyrite (fool's gold) and sulfide compounds have been proposed. At lower, hotter altitudes the material sublimates, but as it rises in the atmosphere it rains down as a sort of metallic snow on the mountaintops of Venus. *Pioneer Venus* also possessed an ultraviolet spectrometer, which took the famous picture of the cloud tops of Venus shown in Figure 2.5. The spacecraft also detected a very weak magnetic field.

The most stunning advances in our scientific view of Venus came from wavelengths that are not within our own biological capabilities to detect. I am often asked why we don't send astronauts out to the celestial bodies we study. I reply that with a spacecraft I feel as if I'm there, and it's a better me that's there because our eyes can't see in the radio, infrared, and ultraviolet regions of the spectrum the same way instruments on a spacecraft can. It's also much more expensive to send people rather than robots into space, and crewed exploration puts lives at risk.

We have sent more probes to Venus than any other planet (and I haven't even mentioned them all), but as 1990 approached we still

FIGURE 2.5 An ultraviolet image of Venus obtained by *Pioneer Venus*, showing its atmospheric cloud structure. NASA.

knew very little about its surface, locked underneath an impenetrable atmosphere. The mapping accomplished by *Pioneer Venus* was not at high enough resolution to give a detailed view of the geologic history of Venus. Most of what we now know about the surface of Venus comes from the *Magellan* spacecraft, which was launched from the Space Shuttle on May 4, 1989 and took data for four years starting in September 1990 in an elliptical orbit around Venus that swooped down to within 183 miles of its surface. Radar stepped forth once more, but this radar was a much better one, capable of taking pictures as well as mapping out the contours of Venus. *Magellan* put together the first clear images of the entire surface of Venus by using the synthetic aperture radar (SAR) technique, which integrates a multitude of backscattered radar signals from a moving spacecraft hitting the same spot. The images show not only geologic features, but roughness and texture.

After *Magellan* put together its altimetry map based on four years of data-taking, NASA celebrated by sending a globe of Venus with the data imprinted on it to all the principal investigators in its Planetary Geology Program. I was one of the lucky people to receive a globe and it still hangs in my family room (Figure 2.6).

FIGURE 2.6 NASA's globe of Venus, showing high areas in brown and low areas in dark blue. See plate section for color version.

Most of the intriguing features of the surface of Venus are visible on the globe, including Ishtar at the North Pole and Aphrodite at the equator, and craters and weird volcanic features and lava flows, but the SAR images are far more spectacular and revealing. Venus's surface is covered with the fingerprints of massive volcanic activity. Giant lava flows such as that in Figure 2.7 coat the planet; the volcano Ammavaru, which is the source of the lava, lies almost 200 miles to the left.

Figure 2.8 shows two types of volcanoes that are unique to Venus. A "pancake dome," similar to shield volcanoes on Earth, which are created by the steady outflow of lava (think Hawaii), appears to the left of center at the top of the image. The one in this image is about 20 miles wide. The large feature in the center of the image is a corona (plural: coronae), which is formed by the upwelling of hot material from the interior of Venus and a subsequent collapse after the material cools. Two or three small pancake domes, and a field of even smaller ones, is superimposed on the large corona, which the IAU has named Fotla Corona. Some volcanoes take on a spider-like

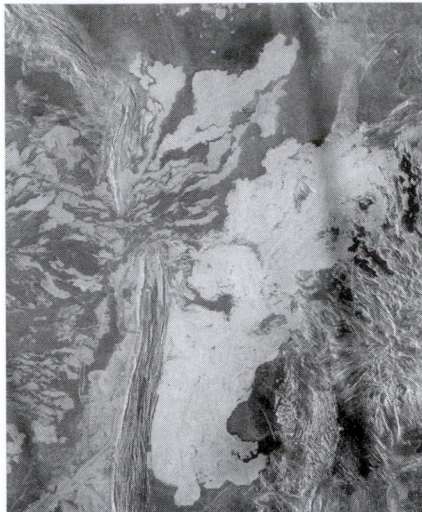

FIGURE 2.7 Massive lava flows on Venus. The image is about 300 miles from top to bottom. NASA/JPL-Caltech.

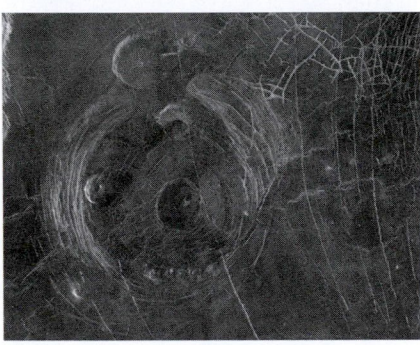

FIGURE 2.8 Two types of Venusian volcanoes. The large one in the center, which is about 120 miles wide, is a large collapsed structure, while the smaller ones are a type of shield volcano. NASA/JPL-Caltech.

appearance and are known as arachnoids; one is even nicknamed "the Tick". Earth-like lava tubes, including the largest one known in the Solar System, are found on Venus. Venus is also criss-crossed by a network of ridges or "tesserae" such as the one left of center in Figure 2.7 that indicate crustal motion and buckling. Clearly, Venus has been an active world.

But things got really strange when planetary geologists working on the *Magellan* project began to count craters on the surface of Venus (several are visible on the globe in Figure 2.6). One reason this tedious process is so important to understanding the evolution of planets is that the number of craters on a surface reveals how old it is, that is, how much time has elapsed since a geologic event (volcanic eruption, flood, etc.) erased all surface features. Older surfaces have more craters because they have been impacted by asteroids from space for a longer time. Most planetary surfaces have terrains of various ages, but the surface of Venus seemed to be all about the same age, 300 to 500 million years old. Was there a massive resurfacing of the planet at this time?

On Earth, volcanoes and mountain building are constantly occurring through the underlying mechanism of plate tectonics. The Earth's crust is in non-stop – but slow – motion, riding on a shell of semi-liquid rock. Crust is being constantly created at the mid-oceanic ridges, and destroyed in the subduction zones where two plates converge (the San Andreas fault is one such zone). It is above these subduction zones that mountain ranges are formed along with associated volcanism. There is no evidence for plate tectonics on Venus: something else drives its geologic activity. One theory is that without energy-releasing plate tectonics, a process that needs water in the interior of the Earth to soften the rock on which the plates glide, Venus builds up heat in its interior and releases it in sporadic episodes of massive planet-wide volcanic activity. The crust cools and the cycle is repeated.

Venus and Earth were birthed in the same region of the inner Solar System, made of the same prenatal material, and poised to continue on similar paths through adolescence and adulthood. So why are they not twins? Why did Venus lose all its water? How do we even know Venus was wet in the past? The answer lies in a chemical subtlety, the amount of "heavy hydrogen" (deuterium in chemists' language) in its atmosphere. Hydrogen usually has one electron circling one proton, but heavy hydrogen has an electron orbiting a proton and a neutron. Water, which consists of an oxygen atom and two hydrogen atoms, is broken up into these two types of atom when it evaporates and is hit by sunlight, and if the temperature is sufficiently hot, the hydrogen absorbs enough energy to escape from the gravitational field of Venus. It is easier to move regular hydrogen than heavy hydrogen, so more of it escapes. By looking at the enrichment of heavy hydrogen, planetary scientists have determined that Venus had about 100,000 times more water in the past. Only small amounts persist in its thick clouds.

Venus lost its water because of a massive runaway greenhouse effect, similar to that starting to heat up the Earth, although it is unlikely Earth will lose all its water as Venus has, at least while the Sun is stable. Water vapor in the early Venusian atmosphere absorbed visible light and re-emitted it as infrared (heat) energy, which was trapped and heated the surface of Venus to enable more water to evaporate. Without water to dissolve and sequester carbon dioxide gas coming from Venus's volcanoes, this gas provided additional heating. The Earth keeps its carbon dioxide gas in check in an intricate and delicate balance. The gas is created by volcanic eruptions, as on Venus, but it is dissolved in the oceans and finds its way into the calcium-rich shells of sea creatures, which die and accumulate on the ocean floor. These ocean floors are gobbled up in the subduction zones of plate tectonics, only to be pushed up and to reappear as mountains and volcanoes spewing out carbon dioxide millions of years later to start the cycle all over again. If the Earth were a little warmer – as Venus may have been as it is closer to the Sun – more water may have evaporated and not provided the buffer for carbon dioxide.

Scientists think that, billions of years ago, the Sun was not luminous enough to keep the Earth's oceans liquid: the Earth had to rely on a weak greenhouse effect to sustain liquid water on its surface. Venus's greenhouse effect did it in, while Earth's helped it along. (Some scientists developed the controversial "snowball Earth" hypothesis, which had the Earth completely frozen one or more times in its history, with life clinging on in the few puddles of subsurface liquid). Or was water's escape due at least in part to Venus's slightly smaller size (5% less in diameter) and density (also 5% lower)? Keeping its liquid water required a delicate balance that Venus was not able to maintain. The current spewing of carbon dioxide into the Earth's atmosphere from the burning of fossil fuels and associated climate change could be so catastrophic to us and other life (the Earth itself will do just fine; Gaia may even be happy to see us gone). Venus most likely did have oceans, lakes, ponds, and streams in the past, but all its water has been lost to space. Since the face of Venus tells us nothing that happened before a few hundred million years ago, we have no idea when it was a steamy planet. And if primitive life arose as it did on Earth, it is difficult to envision its survival, although some astrobiologists look to the thick clouds of Venus as a possible survivors' habitat. But those winds whip around the planet at speeds greater than that of category 5 hurricane (200 to 300 mph, and they seem to be increasing) and contain poisonous – at least to us – sulfuric acid. To add to the drama, lightning is common on Venus.

So the surfaces of the three inner planets show us three different epochs: Mercury's face was formed shortly after the birth of the Solar System 4.65 billion years ago; Venus shows what it looked like a few hundred million years ago; and Earth even more recently (how recently depends on where we are looking). But Venus does not appear to be entirely dead yet. Little hints of activity are taking place even in its current quiescent phase. The sulfuric acid found in its clouds seems to be coming from volcanoes: Venus has an extreme case of acid rain.

In a final chapter to *Magellan*'s exploration of Venus, NASA commanded it to crash into Venus. The European Space Agency has been orbiting its *Venus Express* since 2006, but the United States has not sent a mission to Venus in two decades. Planetary scientists agree that with all the attention on Mars, Venus is overdue for another visit. When I talk about NASA's work with the public, I am often asked why we don't send more missions to the planets. My dry answer is always the same: money. We scientists can think up all sorts of cool Venus missions: rovers and sample collectors, balloons and gliders to study its atmosphere and strange chemistry, and more advanced orbiting radar systems. Just send us the money and we will do all this (and more) for you.

NOTES

1 http://nssdc.gsfc.nasa.gov/planetary/chronology_venus.html

# 3   Mars: The Abode of Life?

On the drive up Interstate 15 from Los Angeles towards Las Vegas, as the land begins to ascend and give entry to the city of infinite dreams and dross, one can take a detour from the journey and turn at the small town of Baker ("Gateway to Death Valley"). The environment that continues north along the scenic California State Route 127 is almost indistinguishable from that of Mars, except for the few straggly plants that hug the road and other places with just a little moisture. Vast ranges of sanded plains greet jagged mountains and dry lake beds. The terrain is so Mars-like, that a mock-up of the Mars Science Laboratory *Curiosity* was tested there in 2012, a year after the actual rover was launched (Figure 3.1). Engineers moved the car-sized vehicle over the desert to see how its wheels would fare in the sandy soil of Mars. *Spirit*, one of JPL's earlier Mars Exploration Rovers, had gone to its final resting place in 2009 after one of its wheels got stuck in the martian sand after more than six years of studying the surface of Mars. The other Rover, *Opportunity*, is still going strong at the time of writing.

The cautious and ingenious engineers at JPL weren't going to let a similar fate overtake *Curiosity*. Their mock-up, with a mass of only about 38% of the real thing to account for the lower gravity on Mars, was beset with challenging situations similar to those expected on Mars. It was driven up and over the sides of sand dunes, and into the deepest, softest sand imaginable. The wheels of *Curiosity* were engraved with giant treads with immense gripping power, and in a flight of flair and fancy, the prototype was engraved with the letters JPL. When NASA Headquarters saw that they banned it from the final model. Under the "one NASA" policy, the entire Agency had to share in the glory. But the clever engineers at JPL struck back: they embedded the Morse code symbols for JPL in the tire, so the

FIGURE 3.1 JPL engineers testing the wheels of the Mars Science Laboratory *Curiosity* in the Dumont Dunes, about 30 miles north of Baker, California. The sandy terrain is similar to the challenging conditions found on the surface of Mars. The wheels are 20 inches wide. NASA/JPL-Caltech. See plate section for color version.

true geeks' code is engraved in endless trails all over the martian sands.

But in spite of the Mars-like terrain found on Earth, early explorations of the real Mars dealt a series of blows to our fantasy of finding a planet just like our own. As a young girl in July of 1965, I remember being glued to our tiny black and white television to await the return of the first pictures of Mars sent back from *Mariner 4*. I had been raised on a hearty diet of science-fiction classics that painted Mars as a hospitable – or at least habitable – place populated by various races of humans or human-like creatures. Robert Heinlein's *Podkayne of Mars* featured the adventures of a space-faring spunky female teenager (what an antidote to the vision of the NASA's Mission Control in the 1960s, with orderly rows of buzz-cut engineers in crisp white shirts and pocket protectors!), and his *Stranger in a Strange Land* described a spiritually superior, non-consumerist culture of an advanced race of

Martians. Ray Bradbury's *Martian Chronicles* serialized the demise of a human civilization on the Red Planet, as Mars is known because of its color, even to the naked eye. One of the stories in the *Chronicles* irked me even back then, before a feminist consciousness had gripped my generation. In the chapter entitled "The Silent Town," a miner left behind by his fellow earthlings goes in search of any other human being, preferably a woman. The male chauvinist guesses that the most likely place for a woman to be is in a beauty salon, and of course that is where he finds her. But this woman is not beautiful: Bradbury caricatures her as an overweight battle ax, pursuing the poor miner, the last man on a dying planet. It was a bad story, but it showed again that even sordid human dramas could play out in outer space.

I was fascinated by another chapter in Bradbury's book, "The Green Morning," which described the overnight greening of Mars from a barren wasteland to a lush forest. This story foretold the rise of the Mars Underground, a somewhat undercover group of NASA and university scientists that arose in the 1970s and 1980s to advocate "terraforming" Mars into an Earth-like planet. The technological challenges are huge, and not even the most optimistic scientist would predict an overnight transformation. I first became aware of the ideas of this group in 1989 when Chris McKay, an astrobiologist at NASA Ames Research Center, gave a talk after receiving the Urey Prize of the American Astronomical Society's Division for Planetary Sciences (DPS). My Cornell mentor Carl Sagan bestowed the award upon McKay, and then modestly stepped away: he was at the height of his celebrity at the time, but he still attended our professional meetings. All the kitchen staff were agog at his presence. It is traditional for recipients of this award, which is given each year to the most promising early-career planetary scientist in the world, to give a scientific address at the annual meeting of the DPS. At the end of McKay's talk, he showed a series of images of Mars being terraformed from the desert planet it is today into a successively more Earthlike world: wetter, warmer, and cloudier. The last slide of the talk illustrated a Mars with oceans, cloud patterns, and green land. From the normally polite audience, there were some derisive snickers, but there were also some oohs and aahs of appreciation.

And of course there was H. G. Wells's *War of the Worlds*, which had residents of New Jersey panicking in the streets when the story was read on the radio by Orson Welles in 1938 on the day before Halloween. As I waited for those pictures from *Mariner 4* to arrive, in the back of my mind there was also the fading paradise on Mars painted by Percival Lowell (1855–1916), the nineteenth and twentieth century American astronomer.

Lowell was a Boston Brahmin, descended from a prominent line of colonial bigwigs. The mill town of Lowell, Massachusetts was named after Francis Cabot Lowell, a family member who was one of the first industrialists to envision the factory as the efficient manufacturing system that was to become characteristic of the capitalism of the twentieth century. By the mid-nineteenth century the city's mills employed thousands of immigrant laborers. The Chinese-speaking, cigar-smoking, Pulitzer prize-winning poet Amy Lowell was Percival's sister, and his brother A. Lawrence Lowell was the president of Harvard from 1909 to 1933. The latter's legacy revolves in part upon his effort to preserve White Anglo-Saxon Protestant (WASP) privilege at Harvard: he famously attempted – and succeeded – to limit the Jewish enrollment to 15%. He also opposed the nomination of Louis Brandeis, the first Jew appointed to the Supreme Court, banned African–Americans from freshman dorms, and purged the campus of gay men, an action that was unearthed only in 2002, and that tragically led many of the men to commit suicide (and was ironic, as Amy was probably gay).[1]

In 1894, Percival decided to devote a large portion of his family's fortune to building an observatory in the cool dry highlands of Flagstaff, Arizona, which was not a state until 1912. He had become interested in astronomy and was drawn to the writings of Camille Flammarion (1842–1925), a French astronomer and spiritualist who believed Mars was inhabited by an intelligent race that had previously communicated with Earth. Flammarion is perhaps best known for his

FIGURE 3.2 Flammarion's iconic image of the seeker.

production of an 1888 wood engraving that depicts a medieval pilgrim – perhaps an avatar for anyone seeking knowledge or enlightenment – penetrating the heavens to behold the inner workings of the cosmos (see Figure 3.2). The meaning of the scene beyond the sky is debatable: among the interpretations are that it includes the merkavah (upper left), the mystical chariot of Ezekial that inspired medieval Jewish kabbalists, or that it is simply a concrete depiction of the orbits of the planets. In any case, astronomers throughout the last century have considered – and we still consider – the print inspirational, part of the lore that underlies astronomy as being a search for the ultimate workings of the universe.

Lowell had also heard of the discovery in 1888 of "canali" on the surface of Mars by the Italian astronomer Giovanni Schiaparelli. The Italian word means "channels," natural bodies of water, but the word was translated as "canals," a word that implies a man-made sluiceway. Soon after Lowell installed the 24-inch refracting telescope built by fellow Bostonian, Alvan Clark, the ace telescope maker of his day, he began to produce drawings depicting a complex system of canals and dots on the planet (Figure 3.3). Lowell believed the canals were conduits bringing water from the martian polar caps – which are clearly visible through a telescope and follow a seasonal pattern similar to the Earth's – to a dry, dying civilization in the temperate zones. The dots at the intersections of the canals were "oases" in the martian desert. Lowell also claimed to see seasons: as the polar

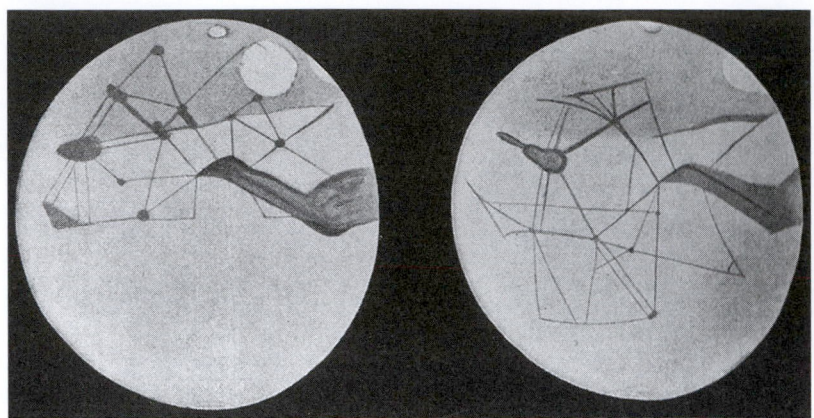

FIGURE 3.3 Drawings by Percival Lowell of the canals, oases, and advancing green spring vegetation on Mars (top).

caps retreated in the spring he could see a blanket of green spreading toward the equator. Lowell published three books describing in detail his passionate beliefs about an advanced civilization on Mars: *Mars* (1895), *Mars and Its Canals* (1906), and *Mars as the Abode of Life* (1908).

Almost as soon as Lowell popularized his ideas of civilized life on Mars, other astronomers attacked his observations and his credibility. Eugene Antoniadi turned the larger 33-inch telescope at Meudon University toward Mars and declared the canals to be optical illusions. Alfred Russel Wallace (1823–1913), the British biologist best known for his independent discovery of the theory of evolution, published a book *Is Mars Habitable?* (1907) that demolished Lowell's claims. Wallace showed that fundamental physical principles precluded the existence of a system of canals and intelligent life on Mars. The planet was much colder than Lowell claimed, and the atmospheric pressure at the surface was too low for water to exist as a liquid. If there were water, it would have been detected in the atmosphere by terrestrial telescopes. When a 60-inch telescope was built on California's Mount Wilson in 1908 and trained on Mars, the canals were shown to be systems of separate dark patches. The human eye tends to organize such patches into linear, continuous structures. Psychologists even have a term for this tendency to see patterns when none are there: apophenia. The canals were indeed optical illusions, and Lowell's observation of vegetation advancing in the spring – if it were real – may have been a large dust storm of the sort we now know periodically envelopes Mars.

Lowell's ideas seem even more ridiculous today, but his critics have overlooked one aspect of his work: he was an early environmentalist, a visionary who harped on about the need for sustainability. *Mars as the Abode of Life* was prescient in its tone about the need to preserve land, water, and resources. He speaks of a nearly global desert on Mars, beautiful but fatal to its inhabitants:

> A vast expanse of arid ground, world-wide in its extent . . . Their bare rock gives them color, from yellow marl through ruddy sandstone to blue slate . . . But this very color, unchanging in its hue, means the extinction of life . . . Pitiless indeed, yet to this condition the Earth itself must come, if it last so long. With steady, if stealthy, stride, Saharas, as we have seen are even now possessing themselves of its surface. It is perhaps not pleasing to learn the manner of our death. But science is concerned only with the fact, and Mars we have to thank for its presentment.[2]

Over a hundred years have passed since Lowell wrote these words, and the deserts have continued their inexorable march across the face of the Earth.

But as I sat in my family's living room in July 1965, I was hoping to see canals and other wondrous sights on Mars. In the popular mind, the idea of canals and Martians had never really vanished. Science-fiction literature considered civilizations and clement environments on Mars as established canon. About this time I read a report in *Life* magazine about a scientist who built a small chamber under martian conditions and was able to keep primitive terrestrial plants and animals alive in it. But *Mariner 4*'s first fuzzy pictures of the martian surface showed craters like those on the surface of the lifeless

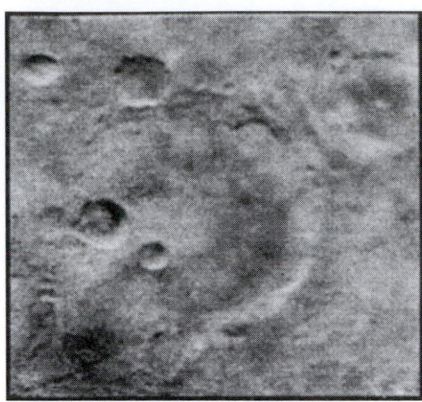

FIGURE 3.4 The best image returned by *Mariner 4*, showing a dead, cratered surface. NASA/JPL-Caltech.

Moon (see Figure 3.4). The march of scientific discovery had peeled away layers of hope and longing, painfully, and decisively. I hadn't been so devastated since the Yankees lost to the Pirates in 1960.

Two more flybys of Mars were executed by *Mariners 6* and *7* (*Mariner 5* flew to Venus), and these craft showed more of the same: craters and desolate landscapes. Our view of Mars as Earth's planetary twin was being deflated even further, just as it had for Venus only a few years earlier. After the failure of *Mariner 8* during launch, *Mariner 9* was successfully inserted into Mars orbit in 1971. It was the first spacecraft to achieve orbit around a planet other than the Earth. *Mariner 9* was also the first in a succession of Mars missions that dug us out from the hole in which we found ourselves after *Mariner 4*. With an orbiting spacecraft, we could finally get a global view of Mars. Except the planet was engulfed in a giant dust storm when it arrived. After the storm cleared, *Mariner 9* revealed a Mars that had once been much wetter, and possibly warmer, in the past. In fact, there was evidence that erosional processes occuring on the Earth, such as flooding, wind erosion, and volcanism, had also sculpted the surface of Mars. Figure 3.5 shows a massive floodplain on Mars that resembles terrestrial features such as the Scablands of eastern Washington State. These massive terrestrial flood plains were formed after natural dams restraining giant bodies of collected glacial water burst at the end of the last ice age 11,000 years ago. And then there is Vallis

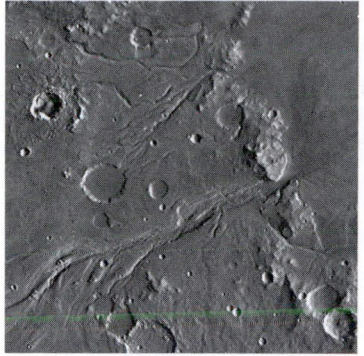 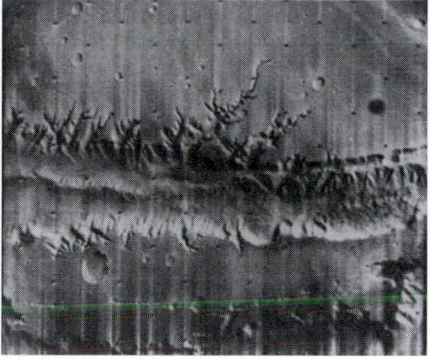

FIGURE 3.5 *Mariner 9*'s discovery of the relics of water erosion on Mars. On the left is a floodplain and associated debris fan, and on the right is the first image of Vallis Marineris obtained by a spacecraft. NASA/JPL.

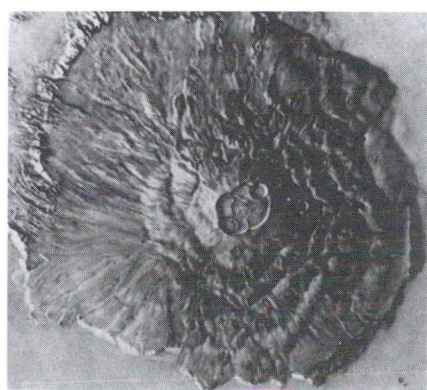

FIGURE 3.6 A *Mariner 9* image of Olympus Mons, the largest volcano on Mars. NASA/JPL-Caltech.

Marineris (Valley of the Mariners): a big gash in the surface of Mars that dwarfs the Grand Canyon and is heavily eroded by running water (Figure 3.5).

We also got a view of pimple-like volcanoes dotting the surface of Mars. Olympus Mons, the largest, is a shield volcano, an immense structure built from successive lava flows (see Figure 3.6). The pancake domes on Venus (Chapter 2) are other examples of this type of volcano. These features arise from conduits filled with molten rock moving up from a planet's interior: they are not part of the super-process of plate tectonics. The best known shield volcanoes on the

Earth comprise the Hawaiian Island Chain. Olympus Mons dwarfs the Big Island – at 17 miles above the mean surface level of Mars, it is more than two and a half times the height of Mauna Kea, measured from the sea floor. Just recently, scientists counted craters – our classic way of figuring out how old surfaces are, as older surfaces have more impact features as they have been exposed to space longer – on the flanks of Olympus Mons to discover that the volcanic activity that created this mountain lasted from about 115 million to just 2 million years ago: a blink of an eye in geologic time. What this discovery means is that Olympus Mons is likely to still be active. It may be a dormant volcano waiting to erupt at any time, a possibility made even more real by the recent discovery of martian methane, a component of volcanic gas.

What Lowell got right was his description of the desert-like conditions on Mars, and now we know the vast sandstorms and dune fields that characterize terrestrial deserts are prevalent on Mars. There are even dust devils on Mars (see Figure 3.7). Planetary geologists have been able to deduce global wind patterns on the surface of Mars from the orientation of its Earthlike dunes. The erosion of the surface of Mars by wind is going on today: the global dust storm that greeted *Mariner 9* on its arrival at Mars is a fairly frequent occurrence.

Another Earthlike feature of Mars is its polar caps, which can be easily observed from the Earth (their melt was the putative liquid that filled Lowell's canals), and which wax and wane with the seasons like those on the Earth. The tilt of Mars's axis is similar to that of the Earth – 25 degrees vs. 23 degrees – and since the tilt of the Earth's axis is what causes the seasons, it isn't surprising that Mars has seasons, too. There are permanent polar caps on Mars, which consist of water ice, and transient caps of carbon dioxide or dry ice (the south polar cap has some permanent dry ice as well), which sublimates in the spring and condenses in the fall. As sunlight creeps up on the ice cap in the spring, violent winds and stormy clouds filled with dust, sand, and even snow sweep up from the surface. The spring heating and release of carbon dioxide is so rapid, that geysers of gas with entrained sand

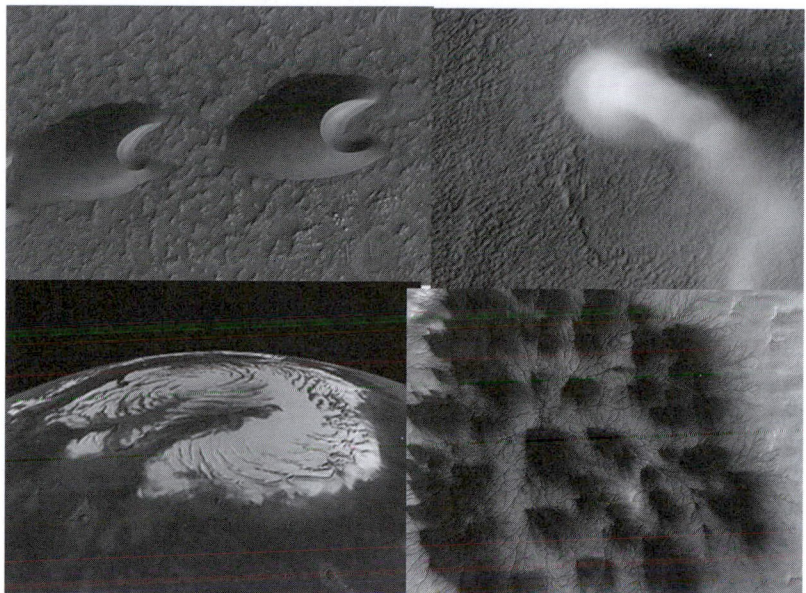

FIGURE 3.7 Marvelous sights on Mars, three of which have terrestrial analogues. Clockwise from the upper left: Barchan dunes, crescent-shaped features common on terrestrial deserts, including California's Kelso dune field in the Mojave desert near Baker, California; a dust devil; spiders on the south polar cap; and the northern polar cap of Mars. Only the spiders are not seen on Earth. NASA/JPL-Caltech/ University of Arizona.

and darker minerals sprout up on the pristine cap to form strange features known as "spiders." Nothing quite like them appears on the Earth, but they make their appearance in the outer Solar System on Triton, as we shall see in Chapter 9.

In 1976, NASA sent two landers, *Viking I* and *Viking II*, to the surface of Mars to determine if life did indeed exist there. When these landers executed their experiments, they showed rather convincingly that there were no organic compounds above the one part per billion level in the upper few centimeters of the surface. The landers also reported on the extreme environment of Mars, with temperatures falling to –120°F (–84°C) at night. So, for a while, the debate subsided surrounding the position of Mars as the abode of life. Mars was seen

FIGURE 3.8 JPL's Tim Parker with his contour map of the northern regions of Mars, showing the extent of a global ancient ocean. Photo by the author. See plate section for color version.

as a desiccated, dry desert, bombarded by ultraviolet radiation. But the spacecraft put into orbit – *Viking Orbiter* – and later missions, continued to show Mars as a past water world.

Ever so slowly, Mars began to wake again. My JPL coworker Tim Parker began to see evidence for global oceans on Mars in the distant past – vast shorelines encircling the smooth basins of ancient bodies of water. He compared remnant shorelines left in the vicinity of Lake Bonneville, the huge prehistoric lake that was the predecessor of the Great Salt Lake, and coastal environments in Antarctica. Clear analogies were there: Mars revealed remnants of debris flows that were deposited above upsloping terrain that looked like the shores of ancient oceans; there was evidence of boulders and rocks pushed by large masses of floating ice; and many craters had been eroded by water on their "seaward" sides. Parker put together a contour map of Mars to show the ancient global ocean covering its northern regions: it was miles deep, comparable to terrestrial oceans (Figure 3.8). Parker

FIGURE 3.9 Sedimentary deposits in the Meridiani Planum region of Mars. NASA/JPL-Caltech/University of Arizona. See plate section for color version.

believes it was sustained by ice-rich impacting bodies brought to Mars during the Late Heavy Bombardment, which ended about 3.8 billion years ago. When the source stopped the water escaped because of the lower gravity on Mars. To quote Parker: "Earth and Venus hogged all the mass, so Mars became just a little bit more interesting than Mercury." As we shall see later, other scientists believe most of the water is still there, trapped in the martian crust. There is also a fair amount trapped in the polar caps: work led by JPL's Jeffrey Plaut showed that all the water in the caps could fill a global ocean nearly 40-feet deep.

Even if the global ocean hypothesis initially aroused some skepticism among scientists, orbiting spacecraft revealed extensive layered deposits, strongly suggesting successive deposits of sediments borne by liquid water. Figure 3.9 shows a spectacular image of such deposits returned by the HiRISE camera on the *Mars Reconnaissance Orbiter*.

Scientists have speculated that life arose on Mars early in its history when it was warmer and liquid water was abundant. Life arose on the Earth almost as soon as it could, right after the subsiding of impacts by space debris during the period of Late Heavy Bombardment that stripped away the Earth's atmosphere and ocean. Mars dwells just outside that sweet spot of celestial mechanics: the "habitable zone" around a star where the temperature is just right for water to exist as liquid. Liquid water is the most important condition for life as we know it: NASA has coined the mantra "follow the water" to describe its search for extraterrestrial life. A planet's distance from its sun is the main factor in determining the temperature of the surface, but there are many others: does it have an atmosphere to hold in and buffer heat? Does it rotate sufficiently fast to prevent cycles of baking and chilling? Is its surface dark or bright (dark surfaces absorb more heat and are thus hotter)? Does its atmosphere possess the heat-trapping greenhouse gases such as carbon dioxide and methane? The latter is what killed Venus's chances of being a tropical paradise: its runaway greenhouse effect has raised its surface temperature from a value very close to that of the Earth's to that of an Italian pizza oven (its slow rotation rate is also a problem). Also critical is a planet's size. Martian gravity is only 38% of the Earth's, so it is much easier for its atmosphere, including water vapor, to escape.

As conditions on Mars deteriorated to become the cold, dry desert of today, could any life forms that arose have "hunkered" down in a protected subsurface environment? Such "hunkerers" are abundant on Earth. Life persists in practically every inclement environment, from the bowels of the Earth – at least two miles down – to very salty, acidic, or boiling water, to the depths of Antarctica's oceans, subsurface lakes, and glaciers, and to the upper atmosphere. The Antarctic's frigid Lake Whillans, which lies under an ice sheet a half a mile thick, is home to thousands of different types of bacteria. Do similar "extremophiles" exist in hidden enclaves on Mars? There is good evidence that much of the liquid that once coursed over the surface of Mars has not even been lost to space but is stored as ice

FIGURE 3.10 A fresh crater on Mars showing ice excavated from beneath the surface. NASA/JPL-Caltech/University of Arizona.

underneath its surface. For example, an image of a fresh crater on Mars obtained in 2008 by the HiRISE camera shows a massive excavation of subsurface ice, a possible habitat for primitive life forms (Figure 3.10). As we have observed more favorable conditions on Mars, possibly even ones that are habitable, the thrust of our research has moved into the search for primitive life forms, either alive or dead, as well as geologic evidence for past standing water.

Another path of revitalization for the search for life on Mars began in 1996 when the late David S. McKay of NASA's Johnson Space Center and his colleagues studied the martian meteorite ALH84001 and found several strands of evidence that pointed to life on Mars early in its history. This meteorite was found in Allan Hills, Antarctica, and the oxygen atoms trapped in its vesicles were identical to those found in the martian atmosphere, as measured by the *Viking* landers. Thousands of years ago, a large impact on Mars sent chunks of rock into a trajectory that eventually brought these pieces to Earth. McKay and his colleagues not only detected chemical traces characteristic of life in the rock, including complex hydrocarbons, and clay and magnetite deposits formed in ways similar to those produced by terrestrial bacteria, but there is what appears to be a fossilized bacterium that had once been on the surface of Mars (Figure 3.11). This intriguing fossil of so-called nanobacteria – creatures that are one-thousandth as thick as a human hair – were found along tiny cracks in the meteorite adjacent to the mineral deposits, cracks that could have supported running water. Each piece of evidence was not strong on its own

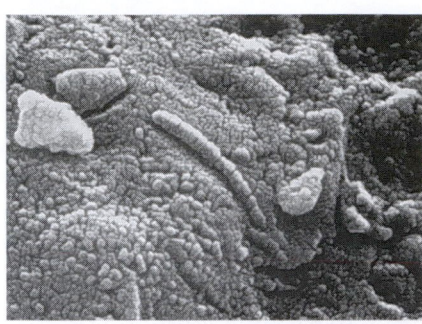

FIGURE 3.11 The martian meteorite ALH84001 showing a putative fossil of a tiny bacterium. NASA.

but, taken together, they provided a compelling argument that this meteorite from Mars once harbored primitive life. Soon on the heels of McKay's announcement, scientists began to criticize every argument. The chemicals were all due to geologic rather than biological processes, or they were terrestrial contaminants, said the naysayers. Furthermore, the nanobacterium – the existence of such creatures is even controversial on Earth – was just a mineral crystal. Even though most planetary scientists ended up on the side of skepticism, McKay's team's discovery energized the field of searching for life on Mars. Again, intelligent speculation and disagreement advanced science.

The series of rovers that trekked over the martian surface provided another key piece of evidence for past habitable environments on Mars: ground truth for sedimentary rocks and minerals that could only have formed in an evaporating ocean or lake bed (Figure 3.12). The tiny *Sojourner* and then the Mars Exploration Rovers, *Spirit* and *Opportunity*, and finally the daringly giant and sophisticated Mars Science Laboratory, *Curiosity* – which dug and delved through the patina of windblown dust that seems to coat everything on Mars – uncovered unmistakable evidence for evaporate deposits: minerals formed in the presence of water. The Mars orbiters, *Mars Odyssey*, *Mars Global Surveyor* (MGS), and the *Mars Reconnaissance Orbiter* (MRO), sent back stunning global views of these same minerals. Carbonates, magnetite, clays, and salts were all there, formed in the same way as they are on Earth, in the presence of water. Moreover, the layered deposits observed by *Curiosity* could only have been formed by

FIGURE 3.12 Extensive sedimentary formations on the surface of Mars shown by the *Curiosity* rover. NASA/JPL-Caltech. See plate section for color version.

sediments set down on the bottom of a lake over tens of millions of years: these lakes were not transient or limited in scope. The driller on *Curiosity* uncovered hematite, a mineral usually formed from an aqueous solution, and which *Odyssey* and *Global Surveyor* showed covered extensive regions on other parts of Mars. Hematite was also found by *Opportunity*, much of it in the shape of "blueberries" or spherules that are formed in solution with water (see Figure 3.13; "Moqui marbles" from the Utah deserts are formed in the same way). The global views provided by the Mars Orbiters, coupled with the ground-truth provided by the Mars Rovers and *Curiosity's* drill, completed the picture. Mars had a sustained watery past. As *Curiosity* Project Scientist, John Grotzinger of Caltech, says about the Gale crater that is the mission's stomping grounds: it is "a system of

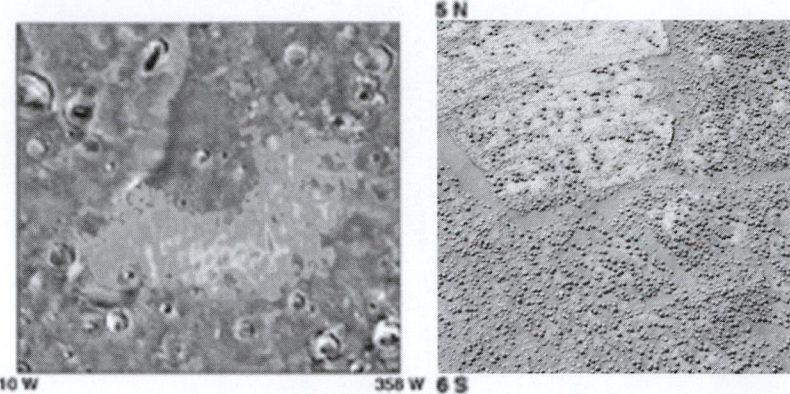

FIGURE 3.13 (Left) An image derived from data returned by the thermal infrared spectrometer (TES) on the *Mars Global Surveyor* showing extensive deposits of hematite, a mineral formed in the presence of water. Higher abundances are hotter colors. (Right) An *Opportunity* image of "blueberries," hematite concretions that are believed to form from watery solutions. NASA/JPL-Caltech/Arizona State University. See plate section for color version.

alluvial fans, deltas, lakes, and dry deserts that alternated probably for millions if not tens of millions of years as a connected system" (*NY Times* Dec. 9, 2014: p. D6). Furthermore, all the elements for a habitable environment were in place: water, energy, and the prebiotic elements of carbon, hydrogen, oxygen, nitrogen, and phosphorus. The only missing element – at least as we know so far – is life itself.

But perhaps even more spectacular than the evidence for standing water early in the history of Mars was the discovery of active gullies on Mars. The HiRISE camera caught an image of deep gullies on the rim of Hale Crater, a typical example of the many found on Mars (Figure 3.14). Some show evidence for fresh and repeated flows, and they look very similar to gullies that form in the California deserts after periodic heavy rains and flash flooding. Scientists originally thought the gullies were formed by running water, as they are on the Earth. Perhaps these features provided the needed evidence for habitable regions near the surface – or even on the surface – of Mars. But the gullies did not form in the warmest regions of Mars:

FIGURE 3.14 A HiRISE image of recent martian gullies in the Hale Crater. The image is 77 × 93 miles. NASA/JPL/University of Arizona.

the ambient temperatures were not nearly high enough to sustain liquid water. Some scientists thought that the gullies were formed by the successive freezing and thawing of carbon dioxide. The march forward to seek habitable environments on Mars was dealt a teasing glimmer followed by a setback. But then scientists found the presence of perchlorate – a salt – at the gullies. This martian salt acts as an antifreeze for subsurface ice, just as salt scattered on wintry roads melts the ice and snow. These gullies are a far cry from the canals of Lowell – but they are still intriguing and speak of a Mars with an active surface and abundant subsurface ice.

One of the most significant lines of investigation for possible life on Mars has been the elusive search for methane. Methane would not last long on the surface of Mars – ultraviolet light would destroy it after only a few years – so if it's there it must be continuously created. There are three possible sources for martian methane: volcanoes (or at least gentler outgassing from below the surface); chemical changes in minerals that produce methane; or life. Early life on Earth did not produce or need oxygen: terrestrial primitive bacteria thrived in oxygen-starved environments and often produced methane as their metabolic output. Thus, if methane is present on Mars, it would provide possible evidence for primitive bacteria similar to those extant on the early Earth. In 2003, two groups reported the presence of trace amounts of methane but, in 2013 the *Curiosity* rover didn't find any at all. Then just a year later, NASA announced the discovery of small amounts of methane (seven parts per billion, on average) on the surface of Mars. The presence of methane on Mars is sporadic! The most pedestrian explanation of chemical reactions in minerals or outgassing may win out, but it is quite possible that methane is a marker for methane-producing bacteria (methanogens, in biologists' lingo) yet to be discovered or, alternatively, active volcanoes. *Curiosity* will continue to monitor the presence of methane to solve this ongoing scientific puzzle.

No discussion of life on Mars would be complete without mention of the "Face on Mars" and its imitators, such as the "Inca City" and the "Airport." In 1976, the *Viking 1* orbiter imaged the Cydonia region on Mars, and a poorly resolved mesa appeared to have rough human features (Figure 3.15). Despite disavowal by all planetary scientists that the face was a relic of an intelligent race, a "Face on Mars" club of gullible fans arose, supported by books and "analysis." Nearby "pyramids" provided further "proof" that the face was the vestige of a past civilization on Mars.

These hoaxes – and more generally all forms of pseudoscience – follow common themes. First, they are sources of profit and fame for those who espouse them, as we saw for Velikovsky's wild claims about Venus. Whole industries have been built around pseudoscientific books and speaking gigs involving the Face. Another usual claim is that NASA and "government scientists" are covering up the truth. Whenever there is an encounter at Mars, a group of nutty protestors shows up at the entrance to JPL, protesting our secrecy. Even if a reasonable person could believe that thousands of maverick, itchy

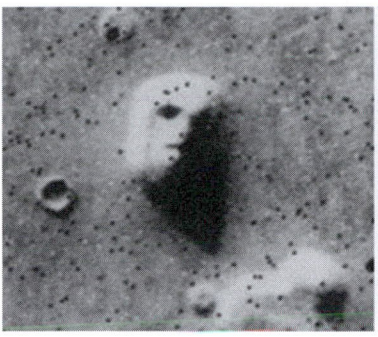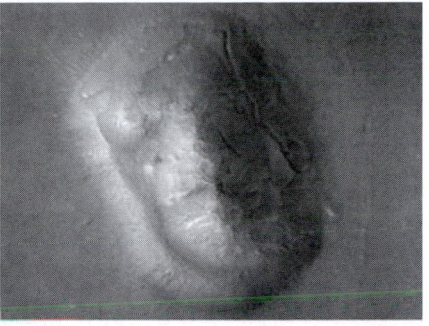

FIGURE 3.15 (Left) *Viking 1* image of a mesa in Cydonia with shadows that suggest human features (the black dots are data drop-outs). (Right) A *Mars Global Surveyor* image of the same mesa under much higher resolution, showing that natural crags and gullies are responsible for the humanoid features. NASA/JPL-Caltech/Malin Space Science Systems.

scientists would shut up, why should we? If we found even slight evidence for extraterrestrial life, our budgets would increase ten-fold, or more. NASA didn't keep the research on the Mars meteorite ALH84001 secret, and in fact it rekindled interest in life on Mars (our budgets increased only slightly). Finally, the scientific method dictates that as evidence is gathered, one must modify models to fit the new data. No amount of evidence will convince the "Face" crowd and their intellectual kin that they are wrong. The *Mars Global Surveyor* obtained images of the "Face" showing it was a hill of eroded gullies and other natural irregularities. But the "Face" enthusiasts persist in their claims.

Other fanciful items found on Mars include an image taken by *Mariner 9* that mission scientists jokingly referred to as the "Inca City," but which seemed to have stuck in the public imagination as a "thing" (Figure 3.16). When scientists looked at the "big picture" they noticed that the "city" was actually a maze of cracks created by a giant impact crater. The most likely explanation for the formation of the city – the "Occam's razor" explanation – is the infilling of cracks by lava and the subsequent wasting away of less resistant surrounding rocks. An "Airport Terminal" (Figure 3.16), also imaged by

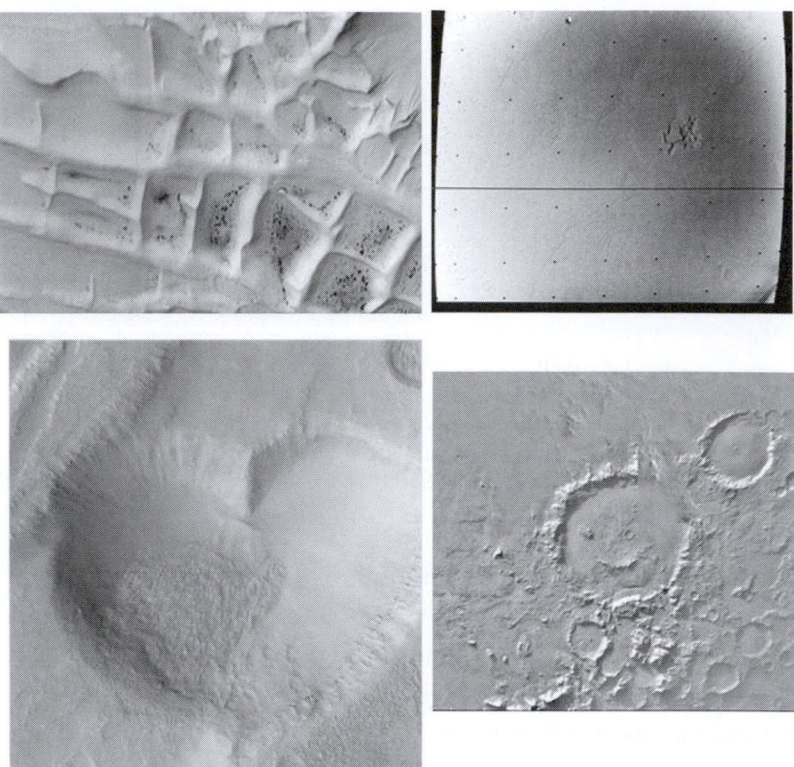

FIGURE 3.16 Fanciful images from Mars. At the top left is the "Inca City," originally observed by *Mariner 9* and shown here through the eyes of the camera on the *Mars Global Surveyor*. The top right shows the original *Mariner 9* image of the "Airport Terminal" (the black dots are fiducial marks, not landing airplanes). At the bottom left is a heart-shaped crater, perhaps formed by an irregularly shaped asteroid, and on the right is a smiley face, with "eyes" and "mouth" formed by smaller craters and ridges. NASA/JPL-Caltech/Malin Space Science Systems.

*Mariner 9*, which its fans claim has runways and parking areas for planes, is a collapsed lava tube or underground cave to the geologist's eye. So far, there are natural explanations for all the exotic features seen on Mars, but a small group of profiteers continues to publish books and give lectures on their obviously incorrect explanations.

Not to be outdone, and to anticipate and preempt some of the crackpot claims – and also to show we do have a sense of fun and romance – NASA scientists have sent out on purpose some of the fanciful features uncovered by spacecraft. For example, on one Valentine's day, NASA sent out an image of a heart-shaped crater on Mars, and later a smiley face (Figure 3.16). There is also a "Mickey Mouse" shaped crater on Mercury. Luckily we weren't sued by Disney, who guards the copyrighted image of Mickey jealously.

Even before the "Face" appeared in the public imagination, the idea of artifacts left by past civilizations on Mars had crept into the common mind. Mars has two moons, Deimos and Phobos ("Fear" and "Terror"), tiny objects that are possibly captured asteroids. In its family of moons, Mars is completely unlike that of the Earth. Phobos (Figure 3.17) orbits around Mars so rapidly that a Martian would see it rise and set three times each martian day, which at 24 hours 37 minutes is very Earthlike. As a 1959 April Fool's joke, the amateur astronomer Walter Scott Houston reported that Dr. Arthur Hayall of the University of the Sierras discovered that both moons of Mars were artificial satellites. (The professor and the school are both fictitious.) He based the claim on a putative decay in their orbits, suggesting they were hollow and thus artificial, possibly a vestige of a past civilization. The respected Soviet astrophysicist, Iosif Shklovsky, who in 1966 published the classic *Intelligent Life in the Universe* with Carl Sagan, didn't get the April Fool's joke – my Russian colleagues tell me the holiday is celebrated in Slavic countries, even during the Soviet period – and produced his own measurements and modeling that supported the hoax, which survived in one form or another for years.

With a little bit of scientific sleuthing, the broad history of Mars has been revealed. Mars even has its own geologic time scale to capture this history in a nutshell. The 4.6 billion year history of the Earth and of all life on it is described by the shorthand of the Geologic Time Scale, a listing of all the ages of our planet (many have passed into common use: pre-Cambrian, Jurassic, Cretaceous, etc.). Although to students on basic geology courses the scale seems like pointless memorization, it is really just a way of holding the world and

FIGURE 3.17 HiRISE image of Phobos, one of the two moons of Mars. It is 14 miles wide (Demos is 7.7 miles wide). The striations coming from the crater, named Stickney after Angeline Stickney Hall, the wife and collaborator of Asaph Hall, who discovered Phobos in 1877, may be related to it, or to tidal forces acting on the moon. NASA/JPL-Caltech/University of Arizona.

all that ever happened on it in the palm of your hand. To a scientist the word "Cretaceous" evokes tropical lands; "Silurian" just sounds like watery seas that were teaming with life, ready to crawl out onto the land; and "Pennsylvanian" had to have occurred when all that coal was formed.

Mars's geologic time scale is sketchier than that of Earth, but it is also a shorthand method to scope out the main train of events that occurred on the planet. As on the Earth, the time periods are named after geographical regions that underwent key events that happened during that period. The earliest events in Mars occurred during the Noachian (named after Noachis Terra – Land of Noah), an appropriate name for a time when the planet harbored abundant water on its surface. The Hesperian period (named after Hesperia Planum, Plain of the West, from one of Schiaparelli's original names) lasted from 3.7 to 3 billion years ago and was characterized by catastrophic flooding and volcanic activity. Finally, the Amazonian (named after Amazonis Planitia – Plain of the Amazons) was a period of increasing quiescence, as the gasping planet lost its water and its volcanoes went to sleep.

But Mars may surprise us yet. There is the teasing evidence that it may harbor extremophiles or other primitive life forms. Its volcanoes may come to life sporadically. Many scientists think that Mars possesses a tremendous amount of water in its crust – estimates range from 6 to 27% of Earth's amount, which is vast if one considers Mars has only 15% of Earth's surface area (but about 50% of its *land* surface area). Over thousands of years, the tilt of the rotational axis of Mars changes drastically – it has no large moon like ours to temper the irregular wobble of its axis – so that sunlight on previously frozen terrain may release ice and even water. But then there are the martian dust storms that dwarf anything on Earth (these storms are probably what Lowell mistook for the advance of green vegetation in the martian spring), the dust devils and dunes, bone-cold frigidity, and an atmosphere that is less than a hundredth that of the Earth's. Will all our teasing evidence for habitable zones, or microbial life, hold up only until the hammer of additional evidence comes down?

We have completed two phases of the exploration of Mars, and we are now on the cusp of the third. The first is that period that every astronomical body goes through: astronomers trying to tediously squeeze information from points of light or blurry disks with fuzzy features open to interpretation. During the current era of space exploration, especially with the rovers, we are studying Mars as a geologic world. We have come full circle: leaving the Mars of science fiction, but returning "to know the place for the first time," in the words of T. S. Eliot. But there will be a third stage in time yet to come: beholding Mars for its mythic beauty. The landforms of Mars look uncannily like those of the Canyonlands of southern Utah, first studied geologically by John Wesley Powell (1834–1902) after the Civil War. Everett Ruess (1914–1934?), the young artist, writer, and explorer who mysteriously disappeared in the Canyonlands in 1934, was the first non-Native person to appreciate this area for its beauty. Writing in a letter to a friend he says "the burning Sun beats down on silent, empty desert . . . long walls of sandstone mesas reach away into the distance, the shadows in their fluted clefts the color of claret. Before me the desert drops sheer away into a vast valley, in which strangely eroded buttes of all delicate and intense shadings of vermilion, orange, and purple, tower."[3] Future explorers will pen similar words as they peer out into the canyons, plains, and valleys of Mars. What we have learned so far places the planet far from the vision of Percival Lowell, but what we find in the future will be more wonderful as we continue our explorations. Poet-explorers of the future will accept Mars on its own terms, not as a "habitable body" fit to conform to our own ideas, but as its very own world. To quote Ruess again, "My burro and I . . . are going on and on, until, sooner or later, we reach the end of the horizon."[4]

NOTES

1 Wright, W. *Harvard's Secret Court*. 2006. St. Martin's Press.
2 Lowell, P. *Mars as the Abode of Life*. 1908. Excerpts from pp. 134–135.
3 Ruess, E. *On Desert Trails*. 2000. Gibbs-Smith Publisher. pp. 45–46
4 Ruess, E. *Ibid*. p. 8

# 4 Asteroids and Comets: Sweat the Small Stuff

There is an apocryphal, but famous, story that has circulated for years among planetary scientists and others. When confronted by two Yale scholars with evidence of stones falling to the Earth after a giant fireball appeared in the sky near Weston, Connecticut in 1807, Thomas Jefferson supposedly said that "It is easier to believe that two Yankee professors would lie than that stones would fall from heaven." The actual comments made by Jefferson were more temperate, reflecting the normal skepticism of any scientist confronted with unusual new data. In a letter dated February 15, 1808 to Daniel Salmon, Jefferson writes:

> We certainly are not to deny whatever we cannot account for. A thousand phenomena present themselves daily which we cannot explain, but where facts are suggested, bearing no analogy with the laws of nature as yet known to us, their verity needs proofs proportioned to their difficulty.[1]

Other historical falls of meteorites include a devastating event in CE 1490, when Chinese records state that stones from the sky weighing several pounds fell "like rain" and killed 10,000 people in what is now Ganzu province.[2] A large tsunami accompanied by fires, which can be caused by an asteroid impacting the oceans, was recorded in both New Zealand and Australia around 1500. (The appearance of Comet C/1490 Y1 and possibly associated yearly meteor showers is probably unrelated as it was observed about four months before the meteor fall.) No surviving meteorites from this event have been found in China. Could it have been a deadly hailstorm with exaggerated casualties? We just don't know.

Even more uncertain is the biblical story of stones that fell on the Amorites: "It happened when they fled before Israel. On the descent of Bet-Horin, God threw upon them large stones (avanim g'dolot) from heaven towards Azekah, and they died. The number that died from the stones of hail (avnai-barad) was greater than the number slain by Israel by the sword" (Joshua 10: 11). The first phrase seems to suggest a meteor shower, but the second phrase explicitly mentions the more common hail. In his book *Rain of Iron and Ice*, planetary scientist John S. Lewis suggests that the halting of the Sun two verses later (Joshua 10: 13) could be an illusion born of a brightly lit sky, similar to that observed after modern, confirmed meteor strikes.[3] Dust in the atmosphere kicked up by the massive impact acts as a highly efficient scatterer.

A handful of events in medieval Europe were reported, but the first terrestrial impact studied in any detail occurred in Siberia. In the morning of June 30, 1908, a column of bluish light turned into a huge fireball and lit up the sky over an area of largely uninhabited taiga near the Tunguska river. Eyewitnesses reported a strong blast following the fireball: S. Semenov was thrown off the porch of a trading post about 40 miles away, as the powerful shock from the explosion swept over the land. Large quantities of light-scattering dust were entrained in the atmosphere over Europe to cause a ghostly illumination in the sky for several nights following the event. Due to the remoteness of Tunguska, and to the turmoil that was gripping Russia at that time, an expedition to the region was not mounted until 1927. A group of investigators from the Soviet Academy of Sciences surveyed the area and took the famous image of thousands of trees flattened by the great blast, all pointing away from the epicenter of the blast (Figure 4.1). The Soviet team interviewed local tribesmen who described a series of four exploding fireballs and explosions.

Rivalling the energy of a hydrogen bomb blast, Tunguska remains the largest confirmed meteor impact in historic times. The impactor, an asteroid, which exploded about five miles above the Earth's surface, appears to have completely disintegrated: only small

FIGURE 4.1 Fallen trees at the location of the Tunguska impact event.

grains of an extraterrestrial origin have been found in the area. If Tunguska had occurred in or near a large city, it would have caused massive devastation and loss of life. Fanciful explanations also abound for the Tunguska event: the explosion of an alien spaceship; the passage of a black hole through the Earth (why was there no catastrophic event on the exit?); an experiment by the moody, increasingly deranged Nicolai Tesla... None of these explanations contains a shred of truth.

The more devastating the event, the more rare. The debris in outer space is dominated by the small stuff. The smallest is the cosmic dust that constantly bathes our planet: every century a layer of this dust equal to about four-thousandths of an inch accretes onto the Earth.[4] That's equivalent to the thin patina of dust collecting on your coffee table after a week or two, though most of that dust comes from other sources such as atmospheric particles and cat dander. The pea-sized impactors burn up in the atmosphere to become the familiar shooting stars, while fist-sized impactors cause a small fireball.

Objects several feet in size have been responsible for the fireballs that appear every year or so, causing some famously (or notoriously) odd but largely non-destructive falls.

In 1992, Michelle Knapp's Chevrolet Malibu was hit by a 26-pound meteorite in Peekskill, New York.[5] When her insurance company failed to reimburse her for this "act of God" she sold the car and responsible meteorite to a collector for $69,000 – a lot of money for an 18 year old. In the evening of January 18, 2010 a fireball and subsequent explosion and smoke trails were reported in the Washington DC area. A few minutes later a baseball-sized meteorite crashed through the roof of the Williamsburg Square Family Practice in Lorton, Virginia and lodged into the carpet of Exam Room #2. The two doctors who ran the office donated the meteorite – which had broken into three fragments – to the Smithsonian for a few thousand dollars, which in turn was donated to the Haiti earthquake relief effort run by Doctors without Borders. In a perfect illustration of how our perception of even celestial events mirrors the prevailing values of society, the landlords of the building claimed they were the true owners of the rock and sued the Smithsonian for possession. Two law students attending William and Mary College and working for the doctors rendered the opinion that the meteorite was similar to lost or abandoned property, which belongs to the finders. The landords dropped their suit, and the Lorton meteorite now resides in the meteorite vault at the Smithsonian.

Perhaps the most suspenseful and ultimately glorious example of an interloper into the Solar System was Shoemaker–Levy 9 (S–L 9). Discovered by Eugene and Caroline Shoemaker and David Levy on March 24, 1993, this disintegrating comet was dead set on a 16-month, inexorable path right into Jupiter. Scientists began taking bets as to what "impact" the comet would have on the planet. I was pretty certain the comet would be swallowed by the thick atmosphere of the great gas giant without a trace. Was I wrong! The 20 some fragments of the comet left a series of fireballs and heat flashes recorded by the *Galileo* spacecraft on its way to Jupiter: their energy was

FIGURE 4.2 A montage showing both hemispheres of Jupiter with the impact scars from the fragments of the S–L 9 comet, below. NASA/European Space Agency/ Space Telescope Science Institute/ MIT.

hundreds of times that of the world's total nuclear arsenal. Prominent dark splotches persisted on the familiar face of Jupiter for months. (See Figure 4.2.)

As many sciences do, astronomy possesses an intricate system of nomenclature that is mostly devoid of science itself. We astronomers cringe when we see school children memorizing the types of stars, or the names and sizes of the moons and planets, as their sole lesson in astronomy. Nevertheless, a few terms make things clearer. Asteroids are the minor bodies that exist mainly in a belt between the orbits of Mars and Jupiter; comets are icy outgassing small bodies; meteoroids are asteroids smaller than one meter (39 inches); meteorites are small bodies that have fallen to the Earth's surface; a meteor or shooting star is a meteoroid as it burns up in the Earth's atmosphere, and a bolide is a large, very bright meteor. Near-Earth objects (NEOs) are asteroids or comets that approach the Sun to within 1.3 astronomical units (AU), and hence to within 0.3 of the Earth's orbit. There are subsets of NEOs that I won't go into: Wikipedia has an excellent article on these. Comets are named after their discoverers (with a P for "periodic" and a number designating the order in which they were discovered) while asteroids get an alphanumeric designation that consists of the year of their discovery and a set of letters and numbers that designate the order in which they were discovered during that year. After their orbits are well defined, asteroids get a number, and finally the discoverer can choose a name that must be approved by the International Astronomical Union (IAU). Generally, asteroids are referenced by both their numbers and names: 1 Ceres; 4 Vesta, 433 Eros, etc.

Back on Earth and about a hundred years after Tunguska, a similar but smaller impact struck during a Siberian morning in February 2013: the Chelyabinsk event. Russian dashcam videos, meant to capture fraudulent car accidents and now replete with colorful expletives, sprouted up on YouTube within minutes. The world was ushered to a ringside seat of this event unfolding: miraculously no one was killed, although about a hundred people were hospitalized and two seriously injured. Heroic teachers rushed their pupils under their desks as if a cold-war style nuclear attack drill was occurring. Whole walls gave way. Windows in the Chelyabinsk Drama Theatre were shattered to leave a carpet of sparkling crushed glass in its entryway. Like Tunguska, the Chelyabinsk house-sized bolide exploded into many fragments miles before it hit the Earth's surface. But this time, hundreds of small stony meteorites were found on the ground, including one dredged from the bottom of Chebarkul Lake that weighed more than a half ton (1,260 pounds).

Further fueling the public's interest in catastrophes from outer space were science fiction and Hollywood. In 1977, Larry Niven and Jerry Pournelle published the dystopian novel, *Lucifer's Hammer*, in which a comet breaks up and slams into the Earth, destroying most human life. The usual ensues: cannibalism, a nuclear war, and power struggles among warring fiefdoms. This novel placed asteroid impacts among the class of bona fide natural disasters in the public mind. A generation later, two movies released in the summer of 1998 about asteroid and comet impacts grabbed the public's attention even more. They were *Deep Impact*, featuring Morgan Freeman playing a cool, measured president who has this calamity thrust at him, and the blockbuster *Armageddon*, which involved a star cast of bungling space travelers and which was deemed less scientific by the astronomical community. In both movies, the deadly invader was destroyed with giant atomic blasts.

Astronomers had known for decades that meteorites – those chunks of rocks from outer space – came from the asteroid belt. Chemically, they didn't match the Earth's composition; they were enriched in rare earth elements for example. Furthermore, their spectra matched spectra obtained by telescopes of the asteroids, implying they had the same composition. One particular class of meteorites, known by geochemists as the basaltic achondrites, is believed to be very similar in composition to the asteroid Vesta, the third largest asteroid. It was a puzzle how these meteorites were transported to Earth.

But the realization that possibly an asteroid or comet caused the extinction of the dinosaurs (except for those that evolved into birds) and most other species of plants and animals 65 million years ago at the interface of the Cretaceous and Tertiary geologic periods was what really started to get scientists, and then the US Congress, alarmed. When first presented by Luis and Walter Alvarez at the University of California at Berkeley in 1980, the idea that an impact – which was dubbed the K–T impact – could cause a mass extinction was roundly rejected by most geologists. They preferred an Earth-based cataclysmic event, such as a giant volcanic eruption, which did in fact occur at about the right time in the Deccan flats in Northern India. Again, a fundamental disagreement among scientists propelled science forward. The key piece of evidence that tilted many scientists toward the astronomical cause was the discovery of a world-wide layer of the rare element iridium. This deposit was laid down in the Earth's stratigraphic record – the successive layers of minerals and elements that tell the history of our planet – about 65 million years ago. There was no plausible terrestrial origin for this rare commodity; the Earth's surface contains only one part per billion of iridium and there is no special place where it's enriched to the extent of the iridium layer. But many asteroids have unusually high abundances of rare elements. The hypothesis was that a giant asteroid or comet struck the Earth and caused massive amounts of impact debris to be lifted and entrained into the atmosphere, and later deposited in a global layer. This dust shut out the Sun for years, stymying the growth of plants, and cutting off the food supply of dinosaurs and most other land species. The impact itself didn't kill off the majority of unlucky reptiles; it was the aftermath of darkness and starvation that did them in.

Many scientists still remained skeptical of the giant impact theory of extinctions until the "smoking gun" was found: an impact crater. Studying the topography and gravitational anomalies on the Yucatan Peninsula, geologists discovered the traces of a crater about 110 miles in diameter that seemed to be the right age. Glassy blebs of minerals fused by heat were also found in the region. With the story complete, scientists began to worry that if the dinosaurs and millions of other species could be wiped out by a giant meteoroid, maybe we could be too.

Several workshops, "white papers," and a peer-reviewed paper published in 1994 by Clark Chapman of Southwest Research Institute and David Morrison of NASA Ames, all emphasized the need to search for and catalogue NEOs. A minority of scientists comprised a second faction believing that an asteroid impact was too unlikely to worry about, and money poured into a search could be better spent on something else. Their argument boils down to this: the Earth is a very small object in space, and space is immense. One way of picturing the sizes involved is to imagine the Sun and planets as fruits of equivalent sizes and placed at equivalent distances. If the Sun is a grapefruit, the Earth would be a poppy seed 17 feet away, and Jupiter is the size of a pea 90 feet away, while Neptune is a mustard seed over a mile distant. The asteroids are mostly invisible pieces of dust floating around.

Nervous about the possibility that, to use the words of JPL's asteroid sleuth Don Yeomans, errant asteroids might "find us before we find them," in 1998 the US Congress finally established the Spaceguard goal of finding 90% of NEOs larger than a kilometer (0.6 mile) – those responsible for serious havoc and worse – within ten years. Scientists believe there are about 1,000 such large objects, and millions of smaller ones. The first NEO, 433 Eros, was found in 1898,

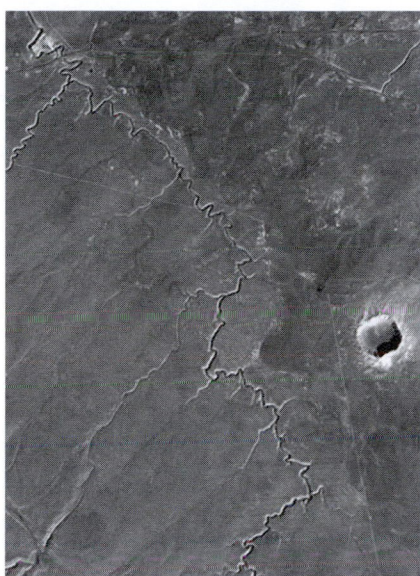

FIGURE 4.3 Landsat image of Meteor Crater, Arizona. NASA.

and a handful were discovered in the first decades of the twentieth century, but this initiative from Congress is what really got the search process moving. Joining the early search pioneers, Eleanor "Glo" Helin of JPL and Eugene and Carolyn Shoemaker at Caltech, were an entire network of astronomers and telescopes, including ones in Arizona, MIT, and New Mexico. As of the end of 2015, nearly 13,504 NEOs have been found.[6] With the success of Spaceguard, Congress and NASA initiated a second program in 2005 to find objects down to 140 meters (359 feet). These bodies are more common, and one would cause crippling regional destruction on the Earth including great loss of life if it landed near a populated area. It is unlikely a bolide would hit a major city, as impact fragments did in *Armageddon* when they destroyed both Shanghai and Paris, because most of the vast expanses of the Earth are uninhabited or sparsely settled. An event such as the one that created the 0.72-mile wide Meteor Crater in Arizona 50,000 years ago (prior to the settlement of humans in North America) was caused by a meteoroid estimated to be 160 feet and happens about every one or two thousand years (Figure 4.3). Scientists believe a

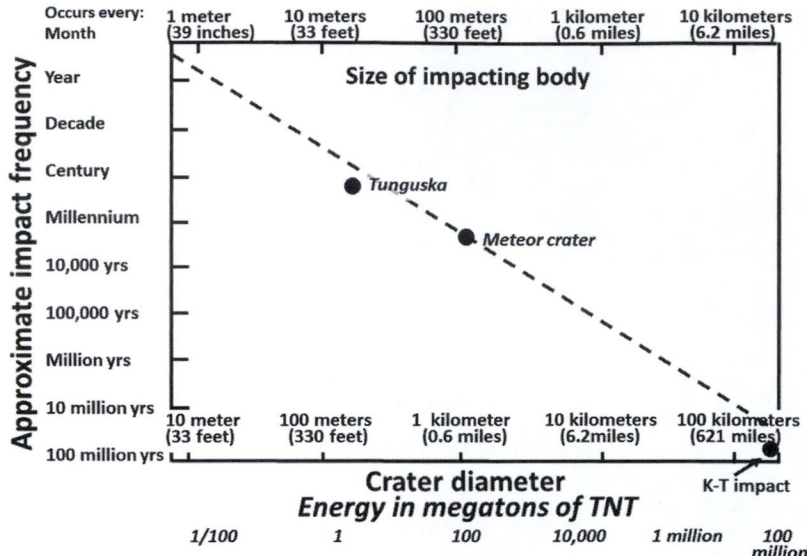

FIGURE 4.4 A summary of our knowledge about the impact frequency and destruction of near-Earth objects. Impactors are likely to burn up in the atmosphere when they are less than about 30 feet. Continent-wide disruption occurs at an impactor size of about 350 feet. Devastation of agriculture and civilization occurs for bodies of about 0.6 miles, and mass extinctions occur for bodies of 6 miles and larger. The dotted line is based on our current knowledge of the flux of NEOs of various sizes. The graph is derived from information in Ward and Brownlee 2000,[7] updated by Paul Weissman of Planetary Science Institute and Alan Harris of MoreData.

Tunguska-scale event happens once every few hundred years or so, although these estimate are fraught with uncertainty.

An entire generation of scientific work is summarized in Figure 4.4, a graph that shows how frequently impacts of various sizes and destructive capabilities occur (the dashed line). But there is a great deal of uncertainty in that graph. Scientists can study the orbit of an individual asteroid or comet and predict with fairly good certainty whether that object will collide with the Earth. But how often will *any* piece of space debris collide with our planet? To answer that question we need to have both an accurate inventory, including the size and number, of all possible objects that could collide with the Earth, as well as their orbits. One good estimate, based on work by Amy Mainzer and her team at JPL, is that we have found 93% of all asteroids larger than one kilometer and none of them represent a near-term threat.

Perhaps an easier method of estimating the number of impacts that occur on the Earth is to count craters on the Earth and Moon. The surface of the Moon is old: if you can count the number of craters of any given size on the surface, and divide by its age deduced from lunar rock samples, you know how often craters of that size were formed on that surface. But this technique is error prone. For one thing, we're not sure how steady the flux of bodies pummeling our corner of the Solar System is. The rate has gradually decreased since the Late Heavy Bombardment, but we're not sure by how much. We don't know if there are sporadic increases in the cratering rate. Making matters worse, parts of the Moon are so saturated with craters that they lie on top of one another, rendering them difficult to count.

One persistent popular myth is that Earth doesn't have too many craters because its atmosphere burns up all the meteors. In fact, only relatively small stuff – objects less than about the volume of a large living room – burn up in the atmosphere. Those breathtaking shooting stars that illuminate the sky like a giant short-lived firefly are objects smaller than the size of a baseball. If a meteoroid is sufficiently large to cause a sizeable crater, it will get through the atmosphere. Our planet shows very few craters because that they are all eroded away by the incessant cycling of water over the face of the Earth, or they are gobbled up by plate tectonics. Thus scientists need to just look at the Moon to understand how often we are pelted. If you can see a crater on the Moon with binoculars – there are at least a dozen such craters – it would have been powerful enough to cause an extinction event such as the one that caused the demise of the dinosaurs. Impact events that large seem to occur about every 100 million years. This interval is also approximately the frequency of mass extinctions, although no other terrestrial impact feature other than the Yucatan crater has been associated with such an event.

More than two thirds of Earth's surface is covered by water, so most terrestrial impacts and their associated fireballs happen over the ocean. A large oceanic event would be no less destructive. Besides the generation of mega-tsunamis, substantial amounts of water would be vaporized and moved into the atmosphere to cause massive rainstorms and climate change. A large event would puncture the crust under the ocean.

The topmost concern on everyone's mind is whether there is any asteroid that is currently on an impact trajectory to Earth. There were certainly some close calls, including a few boulder-sized meteoroids that have approached within 10,000 miles or so of Earth; these objects would just burn up in the atmosphere. One scare inflamed by a media frenzy came in late 1998. Brian Marsden, the astronomer who ran the Minor Planet Center at Harvard's Smithsonian Astrophysical Observatory, said in a press notice that the chances of 1997 XF11, which is about three miles in size, colliding with the Earth was "not entirely out of the question." This claim was based on an International Astronomical Union (IAU) Circular published by JPL's Paul Chodos and Don Yeomans, who computed shortly after its release that the impact probably was in fact near zero. Then their colleague at JPL, asteroid hunter Kenneth Lawrence, pulled out Glo Helin's photographic film archives and found an image of the object obtained with Mount Palomar's 18-inch telescope in 1990. Lawrence carefully measured its position and showed that the asteroid wouldn't hit the Earth by a long shot: a couple of million miles. The world was safe!

The scenario that usually unfolds is the discovery of an object that holds some threat, but as more precise knowledge of its orbit is secured, the threat is retired. For a while, the thousand-foot diameter asteroid 99942 Apophis was our largest worry. After its discovery in 2004, scientists doing follow-up studies of its orbit were alarmed to find out that there was a one in 37 chance that it would impact the Earth on Friday the thirteenth April, 2029. Further observations and the discovery of images obtained prior to 2004 refined its approach on that day to five Earth radii. Another asteroid, 2010 RF12, has a 16% chance of impacting Earth on 5 September, 2095, but it is so small (about 23 feet) that it will probably burn up in the atmosphere. In any case, further observations of it will most likely eliminate its probability of impacting.

But it is the stealth object that has scientists and disaster specialists most worried – the "cosmic sucker punch," in JPL's Don Yeomans's words. We could be surprised by a pristine comet on its first visit to our neighborhood, or by an asteroid with a strange orbit – for example, one that was highly inclined away from the orbits of most other bodies in the Solar System, located in a place we just didn't normally look.

That brings us to the second biggest concern: if we found an object on its way toward a collision with the Earth, could the hand of humanity reach out and crush this mortal enemy? To answer this question, planetary scientists first need to characterize the physical nature of these bodies. How massive are they? What are they made of? How hard will they be to break apart? Are they solid and rigid, or soft and fluffy?

Before NASA established a program to find possible Earth impactors, astronomers had a pretty good idea how they formed and from where they came. We believe the planets and the Sun, and for that matter all the hundreds of solar systems we are now discovering around other stars, formed from great collapsing clouds of gas and dust, as shown by an artist's hand in Figure 4.5. Gravitational forces and collisions caused ever enlarging clumps of material to coalesce. Called planetesimals, these bodies of dirt and ice became the building blocks of the planets. As soon as one became a bit bigger than its neighbors, its greater gravity pulled the others in. But the process was not perfect; many of these embryonic bodies remained alone, remnants of the birth of the planets and the Sun. Astronomers know of three main repositories for this afterbirth: the main asteroid belt between Mars and Jupiter, the Kuiper Belt beyond Neptune (Chapter 9), and the far-out Oort Cloud, the source of many comets that journey into

FIGURE 4.5 An artist's conception of a solar system about 30 million years after its collapse from a cloud of gas and dust. Planetesimals, the building blocks of planets, are shown orbiting the central star: those that didn't accrete into planets became asteroids and comets. NASA/JPL-Caltech/T. Pyle (SSC).

the Solar System. The "small bodies," as they are known as a group, are a treasure trove of scientific information, containing billions of years of memory. But asteroids and comets often escape their native habitats to wander into the inner Solar System, our own home. These are the ones we need to find and worry about.

The second part of the story – the discovery of this debris left over from the formation of the planets – began on New Year's Day 1801, in the vast empty space between Mars and Jupiter where it seemed a planet went missing. The Jesuit priest, Giuseppe Piazzi (1746–1826), discovered in that space what was initially believed to be a seventh planet, the second planet not known to the ancients (the sixth planet, Uranus, was discovered by William Herschel in 1781). In a preview of Pluto's demise as a full planet, this first object discovered in the asteroid belt reigned as a planet for decades. Named 1 Ceres, the International Astronomical Union – the final arbiter of all planetary nomenclature – now deems both Ceres and Pluto dwarf planets, along with three other bodies in the Kuiper Belt at the edge of the planetary

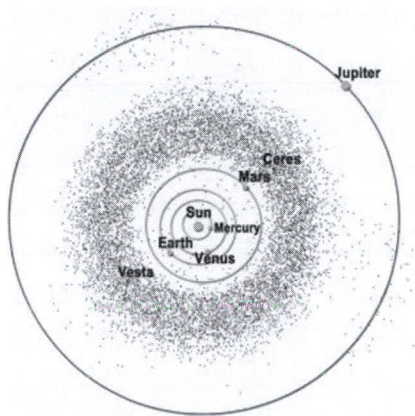

FIGURE 4.6 The asteroids. A small percentage have "leaked" into the inner Solar System. The objects at 11:00 AM and 4:00 PM at Jupiter's orbit are the Trojans. NASA.

system. But, as we shall see, these objects are not in any sense small in their scientific richness or in the stories they tell.

Of millions of asteroids believed to exist, over 600,000 asteroids are catalogued: most reside in the region between Mars and Jupiter, in what astronomers call the Main Belt. But all these objects make up less than 1% of Earth's mass. They exhibit a range of compositions, encompassing stony objects in the inner part of the belt, to carbonaceous bodies residing in the outer part of the belt that are rich in minerals altered by water and inorganic molecules. These compositional differences are imprints of the temperature conditions in the early Solar System. Some are even metal rich, as if they are cores of nascent planets that were stripped of their mantles in violent collisions with other asteroids. Like the Earth, these "protoplanets" differentiated and possessed a metal-rich core. In another indication of the violent history of the Solar System, many asteroids are members of "families": groups of smaller bodies that were formed by the impact and break up of a much larger asteroid. Scientists have been able to tease out the existence of families from their similar compositions, and from their orbits, which can be tracked back to the same place, the point of the impact that formed the family. Another collection of asteroids called the Trojans dwell in gravitationally stable zones in Jupiter's orbit, on either side of the giant planet (see Figure 4.6).

A handful of Trojans are attached to other planets, including one for the Earth, discovered in 2010.

How do asteroids get from the Main Belt into the inner Solar System? Jupiter plays a key role: as the giant sitting close by, its gravity nudges some asteroids into elliptical orbits, which may eventually cross the orbit of the Earth. Speeding the inward journey is a complicated thermal process called the YORP effect (named after the first initials of the four authors of the paper describing the phenomenon), which causes some rotating asteroids to lose energy and spiral inward. Jupiter and YORP work fast (at least by cosmological timescales): NEOs survive for only a few million years before they are swept up by an inner planet or the Sun. But there is a continual supply from the Main Belt.

Still sometimes dismissed as "rocks," asteroids have presented their individuality and troves of scientific information as spacecraft have approached them at ever-closer distances. As it whizzed though the asteroid belt on its way to Jupiter, *Galileo* caught close glimpses of 951 Gaspra in 1991 and 243 Ida in 1993. Ida even had a small moon dubbed Dactyl, laying to rest the scientific controversy as to whether asteroids could be binaries (or more). Now we know about 15% of NEOs have at least one companion; another 15% are loosely bound "contact binaries." In 2000, the NEO 433 Eros became the first asteroid to be orbited by a spacecraft, *NEAR–Shoemaker*.

The connection between Earth and the asteroid belt became superbly palpable when the *Dawn* spacecraft began its detailed investigation of 4 Vesta in 2010. The massive impact basin that showed up even in *Hubble Space Telescope* images came into sharp view: it shows up as the flattened end of the asteroid in the lower left of Figure 4.7. The basaltic achondrite asteroids on the Earth came right from this giant basin. Some were probably flung into fairly direct trajectories, but others arrived many millions of years later as Jupiter and the YORP effect nudged the fragments toward our home. And there are still small asteroids – dubbed vestoids – among the NEO population that one day may impact Earth (none are currently on that path).

FIGURE 4.7 Some asteroids visited by spacecraft. At upper left is 243 Ida and its tiny companion Dactyl to its right; NEO 433 Eros is shown in the upper right; the lower left shows 4 Vesta with its obliterated South Pole – the source of Earth's basaltic achondrites; the lower right shows 1 Ceres with its mysterious bright markings. Asteroids are not to scale; their mean diameters to the nearest mile are (clockwise from Ida): 19; 1; 20; 598; 315. NASA/JPL-Caltech/Johns Hopkins University Applied Physics Laboratory.

Another intriguing appearance of Vesta makes it Earthlike. Planetary Science Institute's Brett Denevi and her team found sponge-like terrain associated with several craters that looks like water or water vapor had bubbled up from underneath. Vesta is a true protoplanet: it went through the early phases of planetary formation that included differentiation into a core, mantle, and crust, and in the process outgassed at least some of its collection of water and possibly other volatiles.

*Dawn* left its orbit around Vesta and embarked on another lonely trek through space for a detailed investigation of Ceres, the very first asteroid discovered. This dwarf planet sent out some teasers in the decades prior to *Dawn*'s visit. In the early 1990s two astronomers, the University of Maryland's Mike A'Hearn and Paul Feldman at Johns Hopkins University, detected the products of water vapor above its surface, while only a year before its encounter by *Dawn* in 2015 two teams of researchers detected an aura of water vapor clinging to its surface. As *Dawn* approached, a mysterious collection of closely spaced, very bright spots came into clear focus on its surface. Were these vents, surrounded by a collar of water frost or deposits left by evaporating water, much as salt is deposited by a disappearing ocean? Did Ceres send up spouts of water from weak locations in its crust, such as craters? And what lurked beneath the surface of Ceres? The bright spots as well as theoretical models of the interior of the dwarf planet suggested a liquid subsurface ocean. Scientists think that terrestrial life arose in the deep oceans, near the thermal vents that sprout up from the abyss and still teem with primitive life today. So the discovery of an ocean on any celestial body leads to the speculation that it may harbor a habitable environment. To be clear, we have not found life, even primitive life, in any location other than the Earth. But we are beginning to discover locations where life could exist, and Ceres appears to be one such place.

One tool in the astronomer's box has opened up another pair of eyes in the study of asteroids: radar, which has solved so many celestial mysteries. Radar astronomers use giant telescopes such as those at Arecibo and Goldstone to bounce radio waves from NEOs (the asteroid has to be close to Earth, otherwise the bounced signal is too weak to detect). The waves are intercepted by the same telescope and constructed into an image that sometimes rivals that from a spacecraft camera. In addition to their sizes, astronomers can also measure the roughness, reflectivity, and other properties of an asteroid from radio waves. My JPL colleagues Lance Benner and Marina Broznovic constructed the image of the binary asteroid 1998 QE2 shown in Figure 4.8.

The Solar System has another belt of debris beyond the orbit of Neptune that was theorized to exist by astronomers shortly

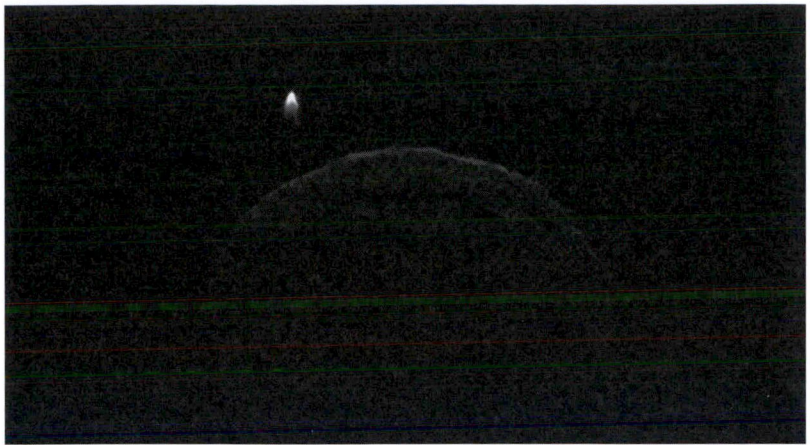

FIGURE 4.8 A radar image of 1998 QE2 obtained at Arecibo Observatory by Lance Benner and Marina Brozovic during its closest approach of 3.6 million miles to Earth in May, 2013. The 1.7 mile wide asteroid has a clear companion, which rotates about the main asteroid every 32 hours. NASA/JPL-Caltech.

after Pluto's discovery in 1930. The first confirmed member of the Edgeworth–Kuiper Belt (named after two of its proponents, and usually called simply the Kuiper Belt) was discovered by David Jewitt and Jane Luu in 1992. Now we know of over 1,500 Kuiper Belt Objects (KBOs), of about an estimated 100,000 total that have sizes larger than about 60 miles. Altogether the mass of the Kuiper Belt is believed to be a small fraction (less than one-tenth) of the Earth's. The objects in this region formed under even colder conditions, so KBOs are made up of frozen ices – water, methane, nitrogen, and carbon dioxide and carbon monoxide, thrown in with rock and a bit of organics (see Chapter 9 for more discussion of the Kuiper Belt).

Astronomers have plotted the orbits of some comets and found from their trajectories that they spend most of their time far distant from the Sun, in a halo of billions of objects that comprise the Oort Cloud, which was first hypothesized by the Dutch astronomer Jan Oort (1900–1992) in 1950. It extends about one third the distance to the nearest star, and we have no way of ever seeing it: the objects it contains are just too distant and faint.

FIGURE 4.9 Image of comet C/2013 UQ4 (Catalina) from the NEOWISE mission on July 7, 2014. NASA/JPL-Caltech/James Bauer.

Comets are perhaps the most beguiling of the small bodies, not only because of the awesome presence they periodically bring to the Earth's inhabitants, but because of their history as harbingers of doom and ruin. In one noted example, the Bayeux Tapestry shows a group of knights pointing ominously to Halley's comet, which appeared in 1066 on the eve of the death of King Harold, the last Anglo-Saxon king before the Norman Conquest. Comets are the pristine building blocks of the Solar System, relatively untouched conglomerates of rock and ice – "dirty snowballs" in the terms of Fred Whipple (1906–2004) of Harvard University, the grandfather of comet science – and my intellectual grandfather as well, as he was the thesis advisor of Joe Veverka, my own thesis advisor. Every so often, the Oort Cloud is perturbed by a passing star or the galactic plane and a comet is set on a path into the inner Solar System. A typical comet is shown in Figure 4.9.

Edmond Halley (1656–1742) was the first astronomer to realize that some comets are periodic. He predicted that the comet that now bears his name would return in 1758, 75 years after its last apparition. The next year, astronomers began calling the first known periodic comet "Halley's Comet" in honor of the late astronomer's prediction. After their passage into the inner Solar System, these interlopers from the Solar System's edges are sometimes captured into orbits around the Sun. (In an interesting historical aside, there is a passage in the Talmud that suggests the early Rabbis knew about periodic comets: "a star appears every 70 years that leads ships astray." Although this

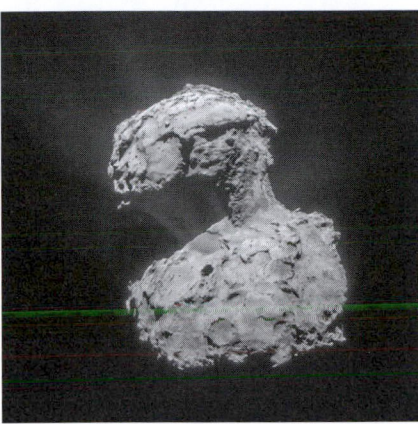

FIGURE 4.10 An image of comet 67P/Churyumov–Gerasimenko obtained by the *Rosetta* Navigation Camera on March 14, 2015 from a distance of 53 miles. European Space Agency.

object could not have been Halley's Comet, it is possible the passage refers to another periodic comet.[8]) The Kuiper Belt is the source of another type of comet: the short-period comets that orbit the Sun with periods of less than 200 years.

The first spacecraft to view a comet were the Soviet's *Vega* and European Space Agency's *Giotto* mission to Halley in 1986. NASA had canceled its own mission to Halley, but over the next generation, the United States sent out a small armada of spacecraft to study comets, and even blasted Tempel 1 with a metallic cannonball. In 2006, *Stardust* returned a tiny sample of comet dust from Wild-2 in its ethereal capture-block of aerogel, a quartz-like substance that is less than 1% as dense as water. But it was the *Rosetta* mission – named after the stone that unlocked the ancient Egyptian language, just as this mission would unlock the secrets of how the Solar System formed – that brought comets into our living rooms. Previous missions had shown flow features, ponds of dust, and holes on comets, but *Rosetta* images of 67P/Churyumov–Gerasimenko showed chasms and cliffs against a backdrop all coated in a layer of dark dust with ice underneath (Figure 4.10).

We know some asteroids turn into comets and some asteroids are dead comets. Several asteroids have been caught "turning on" like a firecracker when pockets of gas collect under their surfaces and explode catastrophically. If there is enough volatile material, the asteroid can become a real comet. On the other hand, some very dark objects appear to be dead comets, the slag-like refuse left over from a life of outgassing. One of the strangest comet–asteroids is 4015 Wilson–Harrington, which showed a glorious tail in 1949 but has been dead ever since. Astronomers now think the tail was the detritus from an asteroid collision and W–H is really an asteroid.

Let's return to the prime question of whether destruction or deflection of a large asteroid or comet moving on a path toward the Earth is possible. My colleague Paul Weissman, now at the Planetary Science Institute, originated the view that comets and many asteroids are essentially rubble piles, conglomerations of small bodies that could possibly be blown apart, as was attempted in *Deep Impact* and *Armageddon*. Other asteroids are rigid, metallic cores of protoplanets. Clearly, we need to study each potential impactor, if we have time. Nuclear weapons are the most powerful exploder, but the Outer Space Treaty forbids the use of such weapons in space. (To save the world it might be okay.) Deflection of the asteroid to a path that does not impact the Earth has also been studied. An especially clever idea is the gravity tractor, a spacecraft planted nearby that exerts a constant gravitational force to alter the orbit of the NEO away from its deadly path. The earlier we detect an object the more time we will have to prepare for what scientists call mitigation, and the more we study these objects with telescopes and robotic missions, the more we will be able to understand their vulnerabilities. Search and characterization are key.

Perhaps the most profound connection between small bodies and us is not about their destructive powers but rather about their life-giving capabilities. Some scientists believe that a substantial amount of water and primitive organic material came from asteroids and comets. The Earth was born when frequent small-body impacts rendered it too hot for water to stick around. When the period of Late Heavy Bombardment eased up, the substance that is the core of life as we know it may have arrived in part from water-rich asteroids and comets that pummeled the Earth early in its history. The details don't quite match: most small bodies have extra "heavy" water, which consists of hydrogen with an extra neutron (but meteorites have the right amount). Complex organic molecules rich in carbon, hydrogen, oxygen, and nitrogen that provide the building blocks of life also formed in the colder reaches of space and made the long journey to the Earth by piggybacking on comets. Some of the water and other molecules in your body was brought to the Earth in this manner.

Astrophysicist Fred Hoyle even had comets as the vectors for panspermia, a theory that life on Earth was brought from the cosmos. Amino acids have been detected on asteroids and comets, but nothing even close to life has been found on any small body (or anywhere else). But still, there is the paradox that the small stuff of the Solar System holds the birth and death of all in one grasp.

NOTES

1 www.monticello.org/site/blog-and-community/posts/who-liar-now A fuller text of the letter is given here.

2 Yau, K., Weissman, P. R., and Yeomans, D. 1994. Meteorite falls in China and some related human casualty events. *Meteoritics* 29, 864–871.

3 Lewis, J. S. 1966. *Rain of Iron and Ice*. Helix Books.

4 Zook, H. A. 2001. "Spacecraft measurements of the cosmic dust flux." In B. Peuker-Ehrenbrink and B. Schmitz (eds.) *Accretion of Extraterrestrial Matter throughout Earth's History*. Springer

5 Yeomans, D. 2013. *Near-Earth Objects*. Princeton.

6 neo.jpl.nasa.gov

7 Ward, P. and Brownlee, D. 2000. *Rare Earth*. Copernicus, Gottingen.

8 Feldman, W. M. 1931. *Rabbinic Mathematics and Astronomy*. Hermon Press New York.

# 5 Galileo's Treasures: Worlds of Fire and Ice

The four main moons surrounding Jupiter have played an outsized role in the history of astronomy, just as Mercury has. They comprise a sort of mini solar system, orbiting in regular orbits about the massive Jupiter. If they were separated from the bright glare of Jupiter, they could all be seen with the naked eye, and there are occasional stories of sharp-eyed individuals being able to see the moons. They range in size from a little smaller than the Moon to larger than Mercury. The four moons seem to have no kinship with each other, spanning the range of celestial personalities from the turbulent Io to the calm Callisto. As is the case with our own Moon, they are tidally evolved and keep the same face toward Jupiter, in a state astronomers call synchronous rotation. In honor of their discoverer, Galileo Galilei, they are called the Galilean moons.

Just as I shall always remember my first view of Mercury, I shall never forget the first time I saw some of Galileo's moons. I was in the fifth grade, and my parents had just bought me a simple little telescope from Hess's, the local department store. It was a reflecting telescope with a 4-inch mirror, magnifying celestial objects only a little bit better than a pair of binoculars. The tube was made of flimsy black cardboard. I eagerly put the telescope together and placed it out on the front lawn. After I secured its wobbly tripod, I pointed it to Jupiter, which was rapidly declining in the west. The Moon wasn't up – otherwise I would have looked at it first – so the sky was dark and Jupiter loomed brightly, seemingly propped up by the trees lining the horizon. Jupiter is the third most luminous object in the night sky, after the Moon and Venus. After some frustrating fiddling, the bright planet lurched into the field of view of my little instrument. I wasn't able to see the Great Red Spot, Jupiter's eternal hurricane, but right next to Jupiter were two tiny dots, pulsating in Earth's turbulent atmosphere and in the scattered light from Jupiter. Ever since this first sighting of two of the four major moons of Jupiter, whenever I view a Galilean moon, whether through a telescope or with a spacecraft, that experience is effortlessly evoked in some deep noetic sense, as Wordsworth's daffodils, which "flash upon that inward eye" to replay the bliss of a past moment.

At that time – the early 1960s – Jupiter had 12 known moons; it now has a staggering 67 companions, and there are undoubtedly more waiting to be discovered. I realized I would be able to see with my telescope at least some of the four main moons that Galileo had discovered in 1610, but I didn't know which moon was which, and it was beyond my grade-school ability to figure out their motions and orbits. I certainly didn't know that the US Naval Observatory published its Nautical Almanac with little diagrams of the positions of the satellites for each night. Of course, now we can just go online and get a diagram of Jupiter and its moons at any time we desire, or download a program that runs the orbits of these moons – or any other moon – in real time. But just looking it up on an app destroys the joy of figuring it out and understanding it all on your own. Working with a computer would never replicate that experience I had with my little telescope.

The Dutch invented the telescope, but Galileo was the first person to use it systematically to study the heavens. His refracting telescope built from two lenses magnified objects about 20 times – the *Hubble Space Telescope* does over 200 times better. Galileo described the phases of Venus (Chapter 2), and the features on the face of the Moon, but his most dramatic discovery was the mini-solar system presented by Jupiter and its moons.

As reported in his very accessible volume *Sidereus Nuncius* (*Starry Messenger*), on January 7, 1610 Galileo noticed "three fixed stars" placed in a line around Jupiter. Over the next few nights these "stars" moved, and one even went behind Jupiter on January 10. A fourth object was spotted on January 13. Galileo rightly concluded

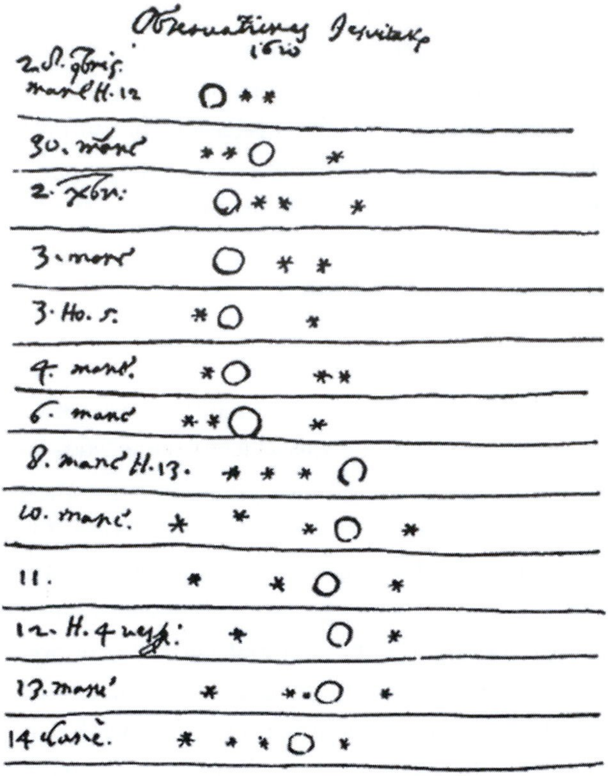

FIGURE 5.1 Drawings from Galileo's *Starry Messenger* showing the star-like four main moons of Jupiter orbiting the planet Jupiter.

that these four moons were orbiting Jupiter, and he was able to compute accurate rotation periods for each of them by plotting their positions night after night (Figure 5.1). Galileo named the four objects the Medician stars, in honor of the Medici family's patronage, but now we know them as the Galilean moons: Io, Europa, Ganymede, and Callisto. Galileo's conclusions directly contradicted the geocentric view of the cosmos, in which all objects orbited a fixed Earth.

Later on in high school, I plotted the positions of the moons from night to night and figured out their periods and distances from Jupiter by using the same Kepler's third law that enabled scientists to figure out the size of the Solar System just by watching the transits of Mercury and Venus (Chapters 1 and 2). But again, this exercise never had the same impact as my first sighting. Galileo had to figure out the orbits of his moons all on his own, as Kepler's third law wasn't published until 1619, and because he didn't have the *a priori* information of the number of moons and their orbital periods. In 2008, Brother Guy Consolmagno SJ, one of my MIT classmates, and the curator of the Vatican's meteorite collection, gave a small group of *Cassini* mission scientists a tour of the Vatican Observatory (which Brother Guy now directs) at the Pope's summer residence in Castel Gandolfo. Soon after being greeted by the words painted on the wall of the observatory, "Deum Creatorem Venite Adoremus" ("Come, Let us Praise God the Creator"), I was able to touch and study a first edition of *Sidereus Nuncius* in the Observatory's library, possibly the same volume studied by Galileo's inquisitors.

Galileo's observations of the orbits of the Galilean moons and the phases of Venus undermined Europe's acceptance of the Ptolemaic geocentric model of the Solar System. From an experimental basis, the geocentric model was adequate: astronomers had planets moving in little circles called epicycles to explain their apparent changes in direction as seen from the Earth. And Church dogma based on the plain meaning of biblical texts pointed to a fixed Earth. But Galileo ascribed to the modern view that biblical descriptions were not to be taken literally; rather they expressed in allegorical form a commentary on the human condition. "The Sun rises and sets and returns to its place" [Eccles. 1:05] is just one example of the rigid interpretation, but the Book of Ecclesiastes was never meant as an astronomical text; rather it is an existential meditation on the futility of all we do. A few sentences later we read the famous sentence that "there is nothing new under the Sun." (In the last chapter we also read that "the making of books increases without end" [Eccles. 12:12].) Over 500 years before Galileo, the great Jewish commentator Rashi (Rabbi Shlomo Yizchaki) wrote that the rising and setting Sun was an allegory of the wicked: even though they rise, at the end, they too will set.

Eventually sentenced by the Inquisition to house arrest for heresy, Galileo was not aware of the treasures his four moons would bestow on his intellectual descendants. And don't forget Nicholas of Cusa (1401–1464) and Giordano Bruno (1548–1600), who went much further than Galileo to posit that each star was a sun with habitable worlds. Bruno earned the *auto de fé* from the Inquisition, although his demise was due to his disavowal of core Catholic doctrines such as the divinity of Jesus and the virginity of Mary rather than to his belief in the plurality of worlds. The final shuttering of geocentrism came with Kepler's laws of planetary motion, which had each planet moving in an elliptical orbit around the Sun, and with Isaac Newton's discovery of the universal law of gravity in 1687, from which Kepler's laws can be derived.

The most mysterious world is Io, named after a priestess of Hera. Like many worlds fantastic, this moon, just a bit larger than our own Moon (2,264 vs. 2,159 miles in diameter), yielded up its first string of mysteries to ground-based observers. With an orbital period around Jupiter of only 1.77 days, Io frequently plunges into Jupiter's shadow to reappear a couple of hours later. Numerous astronomers observed what became known as "post-eclipse brightening": the moon was substantially brighter as it emerged from shadow. Various ideas were bandied about to explain this phenomenon: perhaps a thin atmosphere condensed as Io's surface was cooled by Jupiter's shade; or perhaps Io got irradiated by charged energetic particles from Jupiter. No other moon performed a similar act.

Io also had an unusually reddish color. As early as 1973, planetary scientists noted that its spectrum was similar to that of sulfur, a common element that many of us have played with in our chemistry labs, watching it change from yellow through various shades of orange and brown to a charred black as it is heated up. Io also appeared to be ice free, unlike the other three Galilean moons – and most moons in the outer Solar System – after all, it's cold out there. Io was also bathed in a cloud of sodium. What was that all about?

But the most spectacular claim concerning Io came on the eve of *Voyager 1's* encounter with it. The twin *Voyager 1* and *Voyager 2* spacecraft were launched in 1977 for detailed studies of the jovian and saturnian systems; extended missions carried *Voyager 2* to Uranus and Neptune while *Voyager 1* was sacrificed at Saturn for a close pass to Titan. Stan Peale of the University of California at Santa Barbara and Pat Cassen and Ray Reynolds at NASA Ames published a paper in *Science* on March 2, 1979, just three days before the *Voyager 1* encounter with Io, predicting that the interior of Io was largely molten. They based their calculations of heat in Io's interior on the mean motion resonances among the Galilean moons: for each orbit that Io makes around Jupiter, Europa makes two orbits and Ganymede makes four. This resonance means the moons frequently line up and exert periodic tidal bulges and stresses on each other. As the line-ups – called conjunctions by astronomers – pass, the tidal energy is dissipated as heat. Io moves the fastest, so it has the most conjunctions and thus the most heat and melting. The authors predicted "widespread and recurrent surface volcanism" on Io.

As the first images of Io came in, imaging team scientists were closely scrutinizing them for any evidence of volcanism. Io was a colorful palette of orange, reds, and yellows, with a thin white frosting. There were also numerous volcanic features, but nothing that appeared active. Impact craters were rare, which meant its surface was very young, and the pizza-like appearance of its full disk was "bizarre" and "grotesque," in the words of the scientists (see Figure 5.2).

On Friday March 9, 1979, Linda Morabito, a lead engineer for JPL's optical navigation team, was inspecting an image her team had taken of Io to better refine *Voyager 1's* trajectory after encounter. She was processing the images to enhance background stars and to carefully measure the position of Io with respect to those stars. Predicted and actual positions would enable flight engineers to trim *Voyager's* trajectory to its desired position. Morabito then saw what looked like "a moon behind a moon" (Figure 5.3). A prominent arc-like structure

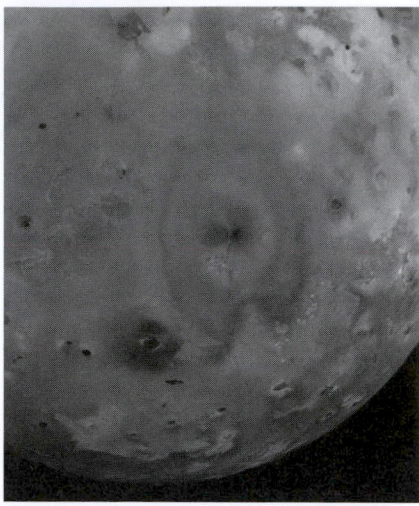

FIGURE 5.2 One of *Voyager 1's* "pizza pictures" of Io, showing the Volcano Pele from a distance of about 244,000 miles. The orange, yellow, and brown hues of sulfur cover the surface. Sulfur dioxide frost is white. Lava flows and crater vents (dark spots) are abundant, but there are no impact craters. NASA/JPL-Caltech. See plate section for color version.

FIGURE 5.3 The *Voyager* navigation team's image of Io after encounter at a distance of nearly 3 million miles. At the lower right of the lit crescent is a volcanic plume from Pele. NASA/JPL-Caltech.

stood out behind the circle of Io. JPL's navigation team's first thought was that a new moon had been discovered, but a quick calculation by Steve Synnott showed that a moon of the size seen would have already been discovered easily by ground-based telescopes. Following the dictum of Occam's razor – the simplest explanation is usually the correct one – Morabito and her colleagues looked for possible instrumental effects: photographic "ghosts" or after-images on *Voyager's* vidicon camera, or perhaps scattered light from some part of the spacecraft.

JPL camera expert Peter Kupferman was called in for a consultation, but none of these explanations could be sustained.

I recently spoke with Linda Morabito Kelly, who now teaches astronomy and mathematics at Victor Valley College in California, about her dramatic discovery. She said the navigation team was exhausted on that Friday, having worked very long hours for two months, but they were still operating at peak performance. "Nothing was going to stop me from finding out what this was; this was exactly the situation I was preparing for." Andy Collins figured out the position of the arc on Io, which the team was now fairly certain was a volcanic plume, and the Project Scientist Edward Stone was called in. Stone immediately agreed that Morabito had discovered a volcano. "This has been an incredible mission" were Stone's words. Morabito was not aware of the Peale et al. article, and she thought the arc was a "satellite behind a satellite" until Stone entered. Morabito and a couple of other scientists stayed late to measure the exact position and shape of the plume. The discovery was announced to the world on Monday morning March 12, 1979, and covered on the front pages of dozens of newspapers. In a dramatic confirmation of a superb scientific prediction, the first example of active volcanism outside the Earth had been found.

Why didn't the imaging team discover the volcanoes that decorate Io? The navigation team had the only image of Io taken after the encounter, looking back at the moon. Like other small particles and dust, volcanic plumes are forward scattering: they are brighter when they are back-lit. Consider the bright light from a movie projector. If you look back into the light, dust in the room becomes illuminated, but if you look into the beam with the light behind you, the dust disappears. The volcanic plume showed up only when it was back-lit by the Sun after *Voyager 1* flew past Io. Morabito's image processing software was also specially designed to enhance faint objects: the stars she was navigating by, like seafarers of old.

But once the science team knew of the plume, additional features that are characteristic of volcanoes popped up. More plumes were discovered. NASA Goddard scientist John Pearl of the infrared spectrometer team detected hot spots on Io, one of which coincided with a plume from Loki, another large active volcano, which appears as the bright spot near the day–night interface in Figure 5.3. A month later, sulfur dioxide was detected in that same plume. *Voyager* data revealed dozens of volcanic features, including a total of nine plumes; many more plume deposits; dark, open volcanic vents filled with molten material (these vents are called calderas by geologists); and deposits of white sulfur dioxide frost from the volcanoes. Robert Nelson of JPL and Bruce Hapke of the University of Pittsburgh wrote a paper showing how the hot hues of Io were all caused by different forms of sulfur – just like our childhood toy chemistry sets.

The *Voyagers* also discovered more than a cloud around Io: a complete torus of glowing ionized sulfur and sulfur dioxide fed by Io's volcanoes coincided with its orbit. From Earth, the cloud appeared as sodium because it is easier to detect with terrestrial telescopes. Plume particles also provided a possible explanation for post-eclipse brightening. Io has a thin transient atmosphere nourished by these particles, which condense as the surface is plunged into the cold darkness caused by an eclipse. This explanation remains controversial. Analyzing *Voyager* images and working with my colleague Joel Mosher, I published a paper in 1995 showing that the night side of Io – which is in darkness far longer at night than during an eclipse – had a thin scrim of condensed sulfur dioxide frost. Other scientists were skeptical of the work, including my own thesis advisor Joseph Veverka, and claimed we were just measuring noise in the camera. A few years later Damon Simonelli at Cornell University led a team that used data from the *Galileo* spacecraft to search for condensation on Io's night side. They found nothing. In my opinion, the jury is still out on what causes post-eclipse brightening – if it is even real – and whether Io's thin atmosphere condenses out at night. Perhaps we see condensation only when the volcanoes are very active and the atmosphere is relatively dense. This uncertainty that always exists at the edge of science is what drives science. Damon was my officemate in graduate school, and I later hired him at JPL. Even though we didn't agree on some scientific issues we could work and prosper together intellectually. I was devastated when Simonelli died suddenly in December 2004 at his home near Pasadena, California at age 45, just as we were finishing a paper together on *Cassini* observations of Saturn's moon Phoebe. Ten years later, the International Astronomical Union was asked by the *New Horizons* Team to name a crater on Pluto Simonelli.

Both subtle and obvious discoveries come from the painstaking months and years of analysis on which scientists embark after receiving a great deal of data. We don't stop after looking at a few images. JPL's Torrence Johnson and his colleagues obtained temperature measurements of the volcanoes using ground-based infrared telescopes to show they were too hot to be mainly sulfur, which has a lower melting point than most rock. As on Earth, these volcanoes had a lot of rock – plain old lava mixed in with just a little sulfur. Sulfur is also brittle: Io's mountains are too high to be composed mainly of it. They must be mainly rock, because a sulfur mountain would collapse.

The next mission to study the jovian system in more detail was the aptly named *Galileo*, which orbited Jupiter between 1993 and 2005. Besides plumes, *Galileo* saw lava lakes and fire fountains (see Figure 5.4), and some mountains higher than Everest. Rosaly Lopes, a scientist on JPL's Near-infrared Mapping Spectrometer Team, counted a total of 71 volcanoes on Io, meriting a record in *The Guinness Book of World Records*. Some of the most crucial observations of Io's activity came from ground-based observations that provided closer temporal coverage. In February 2001, Franck Marchis of the SETI Institute and other observers working at the Keck telescope on Hawaii's Mauna Kea detected the Solar System's biggest volcanic eruption measured in modern times: Io's volcano Surt put out as much power as all of Io's previous eruptions combined, almost 80 trillion watts. This is about four times more power than is expended by all the world's electrical users (recall that power is measured over a specific time period). In six months, the outburst had faded, but *Galileo* detected a reddish

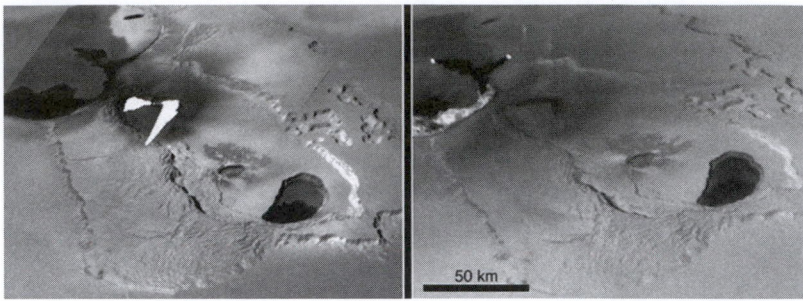

FIGURE 5.4 The volcano Tvashtar as imaged by the *Galileo* spacecraft on November 25, 1999 (left) and on February 22, 2000 (right). The white region in the left image is believed to be a fire fountain that was so bright it saturated the camera's detector. That area was quiescent in the image on the right, but a very active flow, possibly the surface of a lava lake, appeared to its north (left in the picture). The lava flow extending from the central lava lake is a mile high. 50 km is about 30 miles. NASA/JPL-Caltech. See plate section for color version.

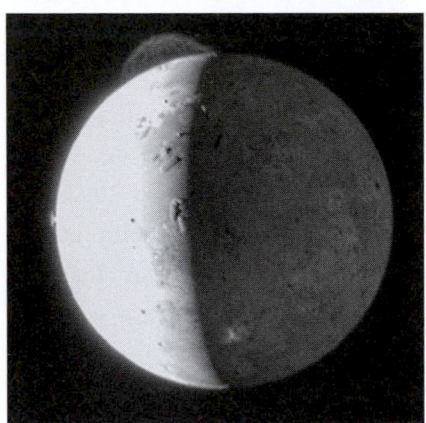

FIGURE 5.5 An image of Io obtained by the *New Horizons* spacecraft on its way to Pluto. Tvashtar is the massive volcano erupting at the top, while Pele is the smaller plume to the left. NASA/Johns Hopkins University Applied Physics Lab.

deposit of fresh sulfur surrounding Surt. In February 2007, during its 1.4 million mile approach to Io, the *New Horizons* mission to Pluto captured a dramatic image of Tvashtar (Figure 5.5) first detected by *Galileo* in 1999 as a mile-high lava mountain next to a liquid silica lake (Figure 5.4). *Cassini* also observed Tvashtar in a state of eruption when it flew by in late 2000. Volcanoes on Io are long-lived, much more so than on Earth. Their eruptions are episodic, as on the Earth, but on shorter time scales.

Io is truly a world fantastic, with a sulfur-laden surface and volcanoes that spew forth sulfur dioxide and also sulfur, most likely as secondary eruption products. Many terrestrial volcanoes also produce sulfur, and sulfur dioxide is one of the most common gases in volcanic vapors. Brennisteinsalda ("sulfur wave" in Icelandic) is a volcano located in southern Iceland that is coated with orange sulfur and surrounded by an aura of sulfur and sulfur dioxide gases. Japan has a number of sulfur-rich volcanoes; Iwo Jima is the English version of Iwo To, which means sulfur island. On Hokkaido, the Shiretoko-Iwo-zan volcano erupted molten sulfur in 1889 and 1936 to form one of the sulfur mountains that abound in Japan.

Science fiction has had a heyday with this fire-and-brimstone world. Writers of the genre had always considered the icy moons of the outer Solar System to be potential habitats for humans when the Sun exploded at the end of its life and left the Earth a burned cinder (don't worry, this won't happen for another 5 billion years or so). But Io presented itself as a modern-day salt mine, the locus for hellish employment, as was depicted in the space western film *Outland*. The best piece has got to be Michael Swanwick's Hugo Award-Winning short story *The Very Pulse of the Machine*, which features the crash on Io of astronaut Martha Kivelsen – whose name is suspiciously similar to UCLA's beloved planetary scientist Margaret "Margie" Kivelson, who did write a few papers on Io, including one in 2011 with UCLA's Krishan Khurana and colleagues entitled "Evidence of a global magma ocean in Io's interior." As the astronaut takes meth to keep her awake while dragging the body of her deceased companion, she hallucinates that Io, with banks and blizzards of sulfur dioxide and sulfur plains, is speaking to her. Is Io alive, in a vision of an expanded Gaia hypothesis, is it a giant alien machine, or is Kivelsen just hallucinating?

The second Galilean moon Europa, which orbits about 417,000 miles from Jupiter, has not given up its secrets so readily. *Voyager 1* approached to within 1.2 million miles of its surface, not quite close enough to reveal a personality (see Figure 5.6). But the closest approach image was intriguing, with faint, long, lineaments

FIGURE 5.6 The best image of Europa from *Voyager 1*, showing a smooth surface with a series of mysterious nearly global cracks. NASA/JPL-Caltech. See plate section for color version.

on its surface. It was also bright, which means fresh. Scientists had four months to speculate on what those strange features were. Since Europa was the first celestial object I studied in any detail as a real scientist, I was paying attention. This was the first time I participated – from the sidelines at least – in the exciting drama of seeing an object come closer and closer, where speculations are confirmed or dashed. I was working on my PhD thesis "Photometric Properties of Europa and the Icy Satellites of Saturn" with Joseph Veverka of Cornell University, a member of the *Voyager* imaging team.

Leading up to a flyby, speculation builds for months, then days, but when the encounter occurs, everything happens very fast as the target bears down. *Voyager 2* approached Europa ten times closer than *Voyager 1* did, and the images that were sent back showed something as odd as Io but totally different. Europa looked like a giant cracked egg. Its surface was smooth and almost crater free, but it was criss-crossed by a series of complex and perplexing lineaments. Scientists began to speculate: Europa had melted and refroze, cracking its surface in the same way a plastic bottle of water cracks if you put it in the freezer; Europa was active; Europa had an ocean . . . We just didn't know. The same tidal forces that kept Io active could also be at work on Europa. The most fascinating features were the triple-band lineaments, which showed a white stripe in the middle, as if fresh ice

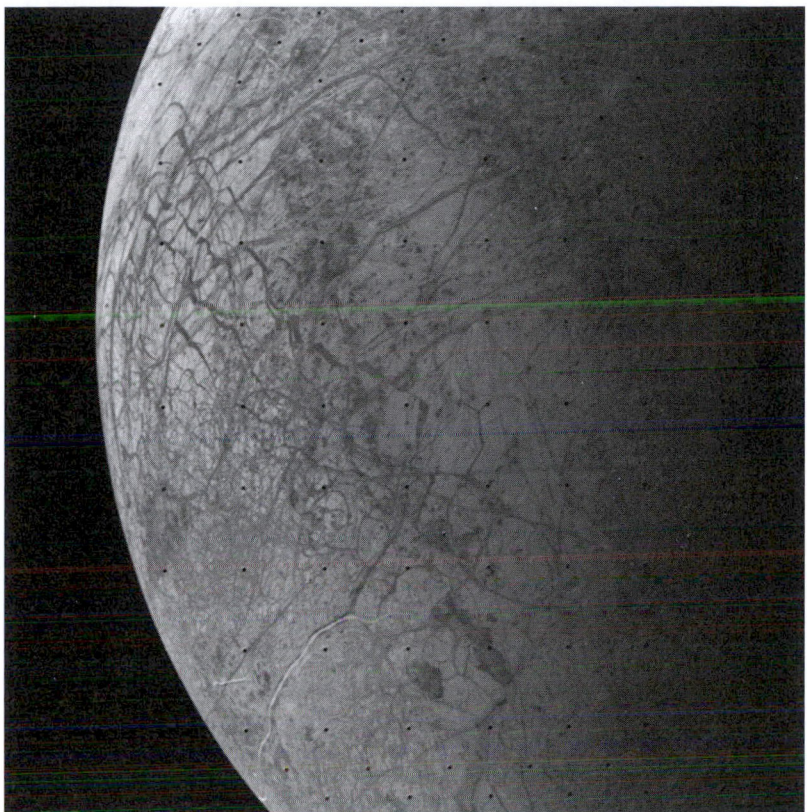

FIGURE 5.7 *Voyager 2*'s image of Europa obtained on July 9, 1979 at a distance of about 140,000 miles. A triple band with a bright center appears near the bottom. Europa has a diameter of about 1,904 miles. The regular small black dots are fiducial marks. NASA/JPL-Caltech.

had been pushing up (Figure 5.7). Europa was also red, contaminated by the sulfur particles that come from Io and drift through space. Io forms a type of ecological zone for Jupiter's inner moons. Amalthea, a small moon inside of Io, is also red.

Voyager 2 took an exit image of Europa back-lit by the Sun, and that image was closely analyzed to search for plumes. None were found after expert analysis, but imaging team member Alan Cook claimed during the 1982 annual meeting of the Division for Planetary Sciences of the American Astronomical Society that the crescent of Europa "exhibited a bright spot" that "extends outside the bright limb [sunlit edge of Europa]." No one believed him, but he was dogged in his pursuit of the evidence. Since he knew I had done my thesis on Europa and had just started a job at JPL as a Postdoctoral Fellow, he asked me to join a paper he had written. I honestly thought the evidence was convincing, and I was flattered that a senior scientist would ask a lowly "postdoc" to be a coauthor. The great Eugene Shoemaker, a grandfather of planetary geology, was another author. (I later became increasingly skeptical of the data, especially after my colleague Mark Showalter, who discovered the saturnian moon Pan in faint *Voyager* images nearly ten years after they were taken, said he could see no plume.) My mentor at JPL was Torrence Johnson, a level-headed, scientifically generous man who later became the *Galileo* Project Scientist. He didn't want to tell me what to do, but he was nevertheless not trying too hard to hide his skepticism. He suggested that I become a coauthor on the paper only if I could demolish the idea of a plume and have the paper come to that conclusion. Haggling over the plume went on and on, and the paper was never published. Again, this tinge of doubt hovered at the edges of science.

Europa seemed to be a transition object between the very active Io and the dead outer moons of Ganymede and Callisto. For my PhD thesis, I studied how scattered light from the surface of Europa can reveal its texture and roughness. I found that Europa, along with the saturnian moon Enceladus, scattered light in a different manner from the other icy moons. Light beams rattled around on their surfaces, giving them a more snow-like texture. Could this "snow" really be great fields of volcanic deposits? As a whole, Europa was also very smooth: no mountains or large crater rims adorned its surface.

Other scientists started to speculate on what might lie underneath Europa's odd surface. In several areas, impacts had punched through the moon's icy crust to cause the movement of rocky slurries of ice over the surface. Pat Cassen showed through theoretical modeling that a liquid ocean could be hiding under the surface of Europa. But the most explosive claim came from my graduate school colleague Steve Squyres and others: Europa was a potentially habitable environment. The current thought on the origin of life on Earth is that it began in deep oceanic thermal vents, nicknamed "smokers." If Europa had a liquid ocean and it was being heated, say by tidal forces similar to those that heated Io, it could harbor a similar habitat. This discussion opened up a whole new world of study: habitable worlds on icy bodies. Scientists began to draw analogies with terrestrial lakes such as Lake Vostok in Antarctica, which lies two and a half miles below an ice sheet, and Lake Whillans, a smaller subglacial lake where microbial life thrives. Could microbial life also exist in Europa's subsurface ocean, a form of extremophile that survives in the coldest, darkest environment? We began dreaming of a mission to Europa to drill through its crust and to find life, the holy grail of planetary exploration. The crust is anywhere from 1 to 20 miles thick – scientists can't agree on the number – and the technology of transporting a drill competent enough for that formidable digging task just costs too much.

But is Europa active at the present time? Intriguing evidence mounted after the *Voyager* encounters and, in retrospect, the most compelling piece of data appears in a paper published in 1989. Observing at NASA's Infrared Telescope Facility (IRTF) on Mauna Kea, Hawaii, Bill Sinton and James Tittemore reported of Europa: "There is only one definite measurement at M [an infrared filter] on 23 April 1981, made at the IRTF. It seems that this measurement, which yields a fourfold increase in the observed flux over any other M measurement, is unassailable. The data acquisition system prints the telescope position encoders, and these values are in good agreement with computed offsets from other Galilean satellites and their encoder readings. Thus, there seems no possibility of misidentification. A total of 25 pairs of integrations were made that were mutually consistent. Filter positions and the amplifier gains are encoded, and there seems no possibility of error in these parameters."[1] What the authors were saying is that they had detected a thermal (heating) event on Europa. This modest finding was tucked into a dense,

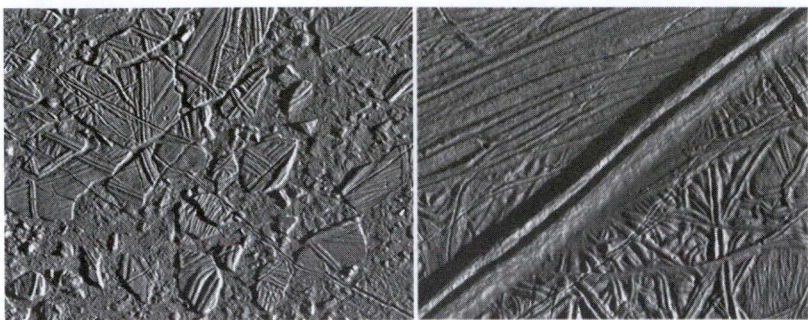

FIGURE 5.8 The left side shows "rafting" on the surface of Europa, in which blocks of ice break off from a major ice sheet that is suspended on top of a liquid ocean. On the right, what appears to be a collar of fine-grain deposit lies around one of Europa's triple band lineaments. NASA/JPL-Caltech.

technical 15-page paper (located behind a paywall, I might add, even though the entire work was funded by NASA), and it was not even mentioned in the abstract. Sadly, both authors have passed away.

*Galileo* flybys of Europa returned even more compelling evidence. Early images showed clear evidence of what geologists call "rafting": blocks of surface ice cracking away from a single solid piece of ice that is floating on an ocean (Figure 5.8). Planetary geologists were even able to put together Europa's rafted pieces like a jigsaw puzzle. Close-up images of the triple bands showed what looked like fine-grained deposits at their edges (Figure 5.8). Some lineaments seemed to be a series of little ice volcanoes, spewing dark material from the subsurface. The darkness of the deposits was beguiling, as organic material – molecules rich in carbon, hydrogen, oxygen, and nitrogen – the building blocks of life, is typically much darker than ice.

Even more speculative was the observation by JPL's Brad Dalton that infrared spectra of sulfur-metabolizing and radiation-resistant bacteria, two groups of terrestrial extremophiles, were similar to the Europan spectra. But nothing remotely suggestive of life, even primitive life, has been found on Europa or any of the habitable

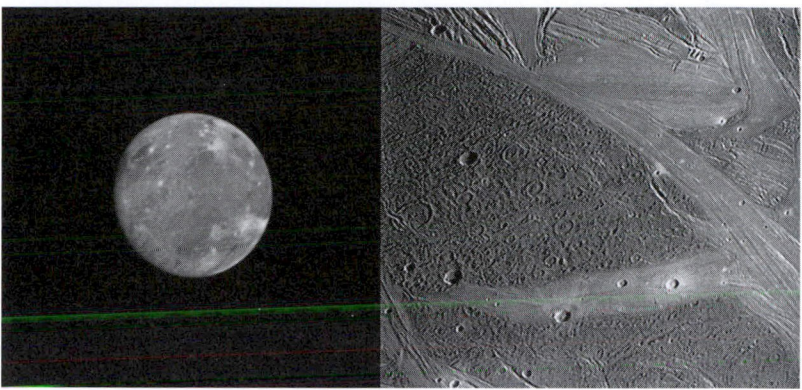

FIGURE 5.9 On the left is a *Voyager* image of Ganymede, showing the dark cratered terrain and the younger grooved terrain. On the right is a *Galileo* image showing clearly the two major surface types. NASA/JPL-Caltech.

environments in the Solar System. Just because life could exist there doesn't mean it does.

All of us who have looked and labored for any signs of an active Europa were finally vindicated in 2014 when Lorenz Roth of Sweden's Royal Institute of Technology and his team, using the *Hubble Space Telescope* (HST), imaged two water vapor plumes 100 miles high. Another possible detection of a plume occurred about two years later. If Europa is volcanically active, it is much quieter than Io. And the type of volcanism is much different: while Io bursts forth with molten rocky lava mixed with some sulfur, Europa's activity and eruptions involve a magma of liquid water, in a process known as cryovolcanism.

Orbiting 665,000 miles from Jupiter, Ganymede was a mixed bag: the *Voyagers* imaged odd grooved terrain (Figure 5.9), but it was dimpled with craters, so it couldn't be the site of current activity. Other parts of Ganymede were much darker and older. The grooved terrain may have formed from a past episode of incomplete tidal heating. Ganymede is about half ice and half rock, and rock is embedded with radioactive atoms that could have provided an extra source of heat early in Ganymede's history. That activity may have just petered out.

The size of Ganymede is one thing that sets it apart: it is larger than Mercury. Not only is Ganymede planet sized (3,273 miles in diameter): it behaves like a planet in many respects. Repeated passages by *Galileo* determined that it had a subsurface ocean and its own magnetic field, a very Earthlike property that only Ganymede possesses among the moons. We believe magnetic fields on planets are caused by a core of a rotating conducting fluid (it's iron on the Earth), which creates a changing electric field and thus an induced magnetic field. Physicists call this effect Faraday induction. Ganymede is thus differentiated into a core, mantle, and crust, just like the Earth. It has the phenomena associated with a magnetic field: aurorae and a magnetosphere, a complex structure of charged particles entrained in magnetic-field lines that offer a shield of protection to the surface below. Aurorae (the Northern or Southern Lights on Earth) form when energy from charged particles trapped in magnetic fields is released as light. On one major measure Ganymede fails as a planet: it doesn't have a substantial atmosphere, just a thin scrim of oxygen and ozone. But because Ganymede has a liquid ocean, it contains a habitable environment, like Europa.

On its final embrace of the Galilean moons, *Voyager 2* sped by the outer moon Callisto in July 1979 to begin its two-year trek out to Saturn. Orbiting 1,170,000 miles from Jupiter and with a diameter of 2,996 miles, Callisto is just short of Mercury in size. Four months earlier, *Voyager 1* had viewed Callisto from an even closer vantage point (Figure 5.10). Finally, here was a moon that was more like what we expected: a dead, heavily cratered world, that had escaped the activity caused by tidal warming. One large impact basin named Valhalla had a series of rings around it – some scientists could count over a dozen – that formed in a fashion similar to the rings that appear when a stone is dropped into a pool. Craters were so thickly placed that Callisto's surface was saturated with them: continuing impacts would just obliterate craters that were already there.

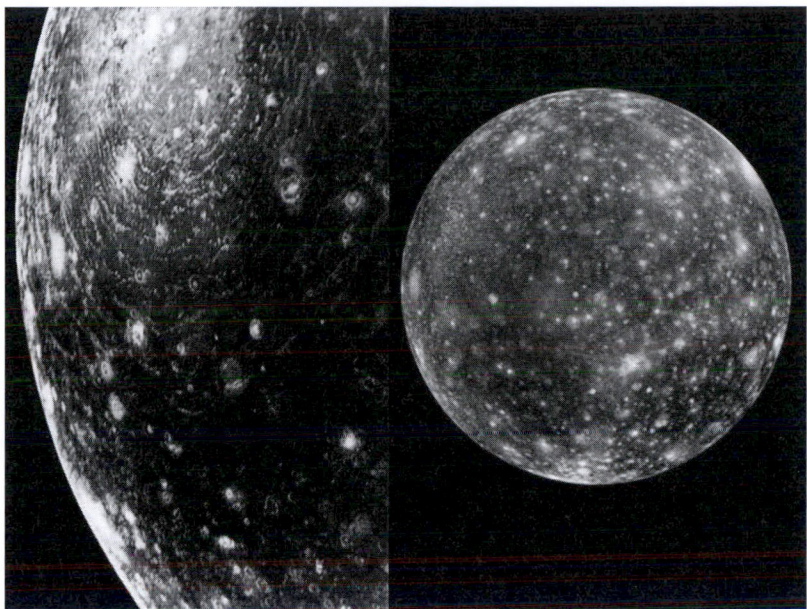

FIGURE 5.10 The image on the left is the closest *Voyager 1* image of Callisto, obtained at a distance of 213,000 miles and showing its dark, heavily cratered surface and the Valhalla impact basin. On the right is a global view from the *Galileo* spacecraft; Valhalla is on the other side. NASA/JPL-Caltech.

In a sense, Callisto was more than we expected, as on the eve of the *Voyager* encounters, planetary scientists were still debating whether craters formed in ice would persist or gradually slump away. Not only did the surface of Callisto show that they persisted, but their shapes and overall appearance were like the craters that formed on rocky bodies such as the Moon and Mercury. It is so cold in the outer Solar System – about –260 °F (–162 °C) on the surface of Callisto, for example – that ice acts like a rock. Callisto is fairly dark for an icy body, reflecting about 20% of the visible light that falls on it. This number is consistent with an old surface that has never seen geologic activity since its birth. Callisto is, in fact, among the oldest surfaces in the Solar System. Thus the degree of geologic activity among the Galilean moons declines with distance from Jupiter.

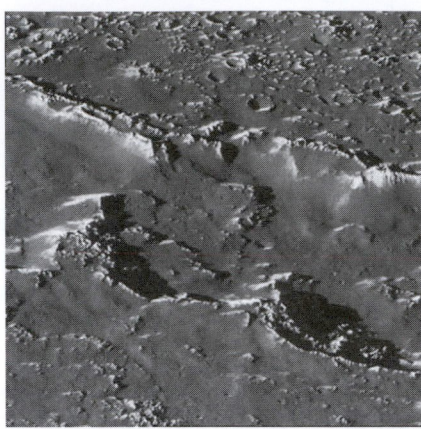

FIGURE 5.11 What appears to be dark accreted dust deposited on top of an array of craters on Callisto. NASA/JPL-Caltech.

But closer up images of Callisto by the *Galileo* spacecraft showed there was still much to captivate and ponder. Many of the crater rims were covered in what appeared to be dusty deposits, and small craters seemed to be absent (Figure 5.11). Jeffrey Moore of NASA Ames and his colleagues hypothesized that this dust was the remains from eons of sublimating ice. Dark material entrained in water ice does not sublimate with the ice; it remains as an inert deposit left behind.

Jeffrey Bell of the University of Hawaii had another idea: he suggested Callisto's "leading" hemisphere, the half that leads the moon's motion around Jupiter, accretes dust as it orbits. This mechanism darkens the leading side of Saturn's moon Iapetus (see Chapter 7), and Bell was saying the outer main moon of Jupiter was scooping up dust around Jupiter in the same way. I published a paper with Joel Mosher of JPL in 1991 claiming that Iapetus, Callisto, and the Uranian moon Oberon were all coated with dark dust originating in the outer tiny moons of the giant planets: this dust spiraled inward and was scooped up by the first main moon. We even predicted the existence of the outer moons of Uranus before they were discovered. The idea gained favor after the discovery of the Phoebe ring of dust around Saturn. So far, no one has found similar rings around the other gas giants, but they may have escaped detection because they are just very faint.

Finally, William Bottke of the Southwest Research Institute and his colleagues published a paper in 2013 entitled "Black rain" in which they fleshed out the dust-deposit model for Callisto, as well as for the other Galilean moons, and showed it was dynamically possible.

Jupiter's magnetic field is conducted around Callisto, which suggests there may be a liquid ocean underneath its crust as well. Thus three out of the four Galilean moons may contain oceans of water that are potential habitable zones, intriguing locales that beg for further exploration.

NOTES

1 Tittemore, W. C. and Sinton, W. 1989. Near-infrared photometry of the Galilean satellites. *Icarus* 77, 82–97.

# 6 Enceladus: An Active Iceball in Space

At about 6:00 PM on a Friday afternoon in January 2005, I was sitting in the waning light of my JPL office and daydreaming about my activities for the weekend. I was startled by the sudden sharp ring of the phone, which I answered promptly – all the travel audit people and administrators wanting a favor done would have left. On the other end was Trina Ray (see Figure 6.1), a crack *Cassini* science planning engineer who was in charge of leading the plans for *Cassini*'s close flybys of Titan. She was sitting in the office of Brian Paczkowski, another fabled JPL engineer who led the science planning office, and they were poring over an image of Enceladus, a small icy moon of Enceladus, that had been taken in November 2004.

"Bonnie, let me email you this image of Enceladus. We've been looking closely at an enhanced version of it. It looks like it has a plume," said Trina.

"Okay, send it over," I said.

I downloaded and looked at the image and sure enough, there appeared to be a tenuous, but well-formed, plume about the size of the crescent moon itself, which is a little over 300 miles from end to end (Figure 6.2). For comparison, our own Moon is 2,159 miles in diameter – you could fit 373 Enceladuses inside the Moon. Even though it was past 9:00 PM on the East Coast, I immediately called Paul Helfenstein of Cornell University, the Imaging Science Subsystem (ISS) representative of the group I was leading to plan all the *Cassini* targeted flybys of Enceladus and the other saturnian satellites.

"No, Bonnie, we are pretty sure that isn't a plume. We've looked at the image, and saw the 'plume,' but we think it's scattered light from the camera. But we're going to do some more tests to make sure." I didn't know it at the time, but Paul immediately called other

FIGURE 1.1 Jupiter and the crescent Moon are the bright objects in the sky, with Mercury just visible above the haze along the horizon (to the right of the leftmost small tree). Image by Steve Edberg.

FIGURE 1.8 A *MESSENGER* image of the Praxiteles crater on Mercury, one region identified by Dave Blewett and his colleagues as a possible outgassing site. The hollow identified by the arrow is a possible vent surrounded by a bright deposit that may be volcanic ash. The crater is about 113 miles in diameter. NASA/Johns Hopkins University Applied Physics Laboratory/Carnegie Institute of Washington.

FIGURE 2.2 Venus over the Pacific Ocean. Image courtesy of NASA. Photograph taken by Mina Zinkova.

FIGURE 2.6 NASA's globe of Venus, showing high areas in brown and low areas in dark blue.

FIGURE 3.1 JPL engineers testing the wheels of the Mars Science Laboratory *Curiosity* in the Dumont Dunes, about 30 miles north of Baker, California. The sandy terrain is similar to the challenging conditions found on the surface of Mars. The wheels are 20 inches wide. NASA/JPL-Caltech.

FIGURE 3.8 JPL's Tim Parker with his contour map of the northern regions of Mars, showing the extent of a global ancient ocean. Photo by the author.

FIGURE 3.9 Sedimentary deposits in the Meridiani Planum region of Mars. NASA/JPL-Caltech/University of Arizona.

FIGURE 3.12 Extensive sedimentary formations on the surface of Mars shown by the *Curiosity* rover. NASA/JPL-Caltech.

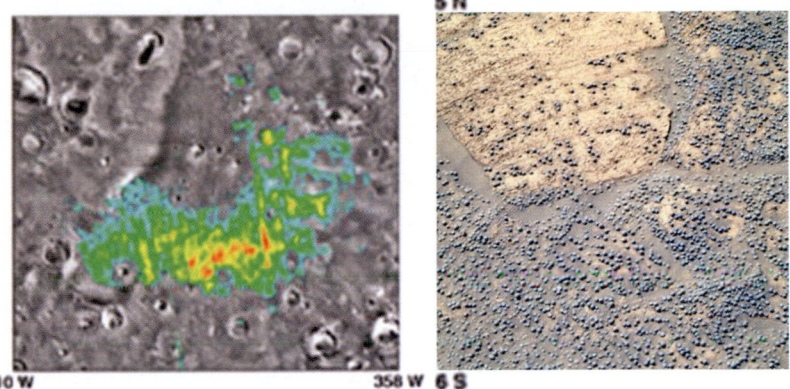

FIGURE 3.13 (Left) An image derived from data returned by the thermal infrared spectrometer (TES) on the *Mars Global Surveyor* showing extensive deposits of hematite, a mineral found in the presence of water. Higher abundances are hotter colors. (Right) An *Opportunity* image of "blueberries," hematite concretions that are believed to form from watery solutions. NASA/JPL-Caltech/Arizona State University.

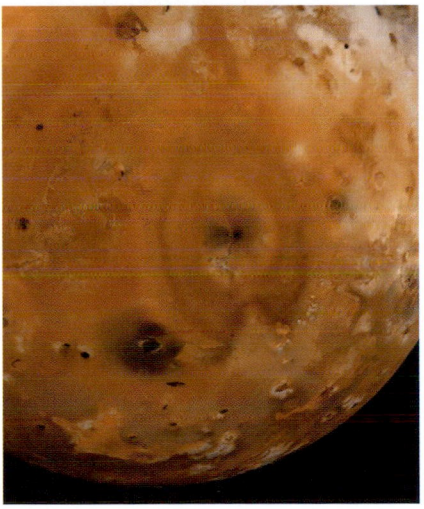

FIGURE 5.2 One of *Voyager 1's* "pizza pictures" of Io, showing the Volcano Pele from a distance of about 244,000 miles. The orange, yellow, and brown hues of sulfur cover the surface. Sulfur dioxide frost is white. Lava flows and crater vents (dark spots) are abundant, but there are no impact craters. NASA/JPL-Caltech.

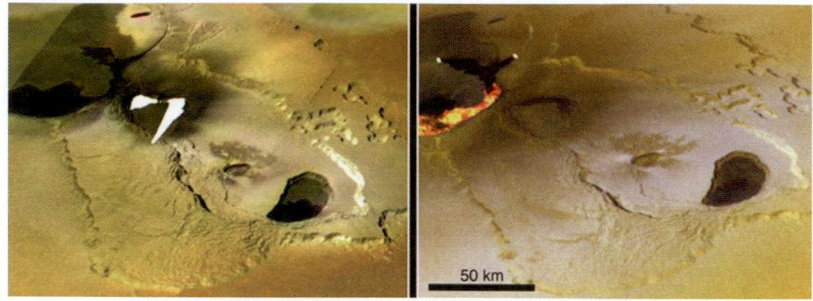

FIGURE 5.4 The volcano Tvashtar as imaged by the *Galileo* spacecraft on November 25, 1999 (left) and on February 22, 2000 (right). The white region in the left image is believed to be a fire fountain that was so bright it saturated the camera's detector. That area was quiescent in the image on the right, but a very active flow, possibly the surface of a lava lake, appeared to its north (left in the picture). The lava flow extending from the central lava lake is a mile high. 50 km is about 30 miles. NASA/JPL-Caltech.

FIGURE 5.6 The best image of Europa from *Voyager 1*, showing a smooth surface with a series of mysterious nearly global cracks. NASA/JPL-Caltech.

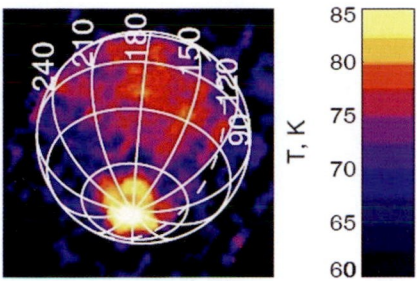

FIGURE 6.8 John Spencer's infrared image showing a large heat anomaly in the southern hemisphere of Enceladus. Courtesy John Spencer.

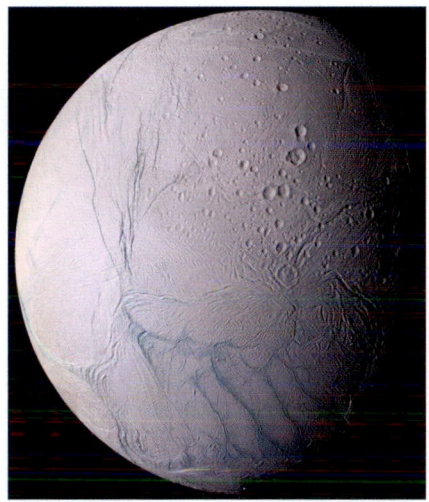

FIGURE 6.10 (Left). The tiger stripes in the south pole of Enceladus are named after cities in the Middle East: (left to right) Damascus, Baghdad, Cairo, and Alexandria. NASA/JPL-Caltech.

FIGURE 7.3 Kim Tryka and Michael Hicks are standing on the far side of the large lawn in JPL's mall areas. They are just dots (if you know where to look). Bringing them 100 times closer transforms them from dots into real people, analogous to the way spacecraft transform a dot in the sky into a geologic world. Photos by the author.

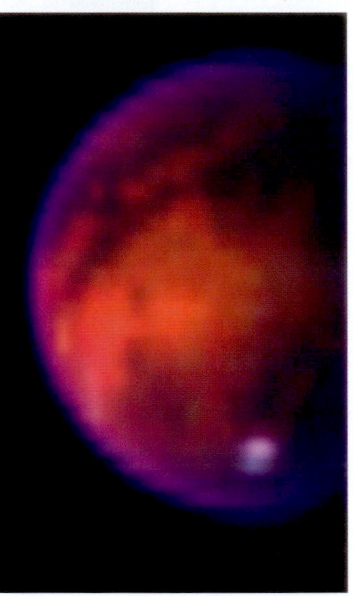

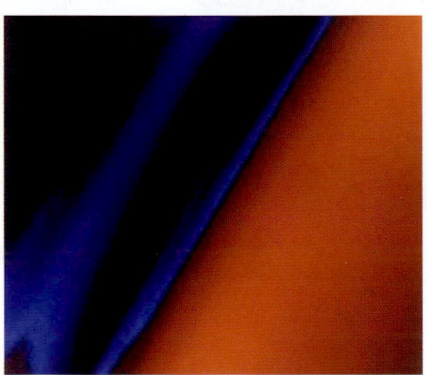

FIGURE 7.4 One of the closest images obtained by *Voyager 1* of the orange atmosphere of Titan and the thick smoggy layer of haze extending out into space. The image has been enhanced in color to show the haze layer and the reddish atmosphere of Titan. NASA/JPL-Caltech.

FIGURE 7.6 Our "sneak peek" of Titan in 2004. NASA/JPL-Caltech/ University of Arizona. Image processing by Tom Momary.

FIGURE 7.8 The surface of Titan as seen by the *Huygens* lander. Note the rounded rocks. ESA.

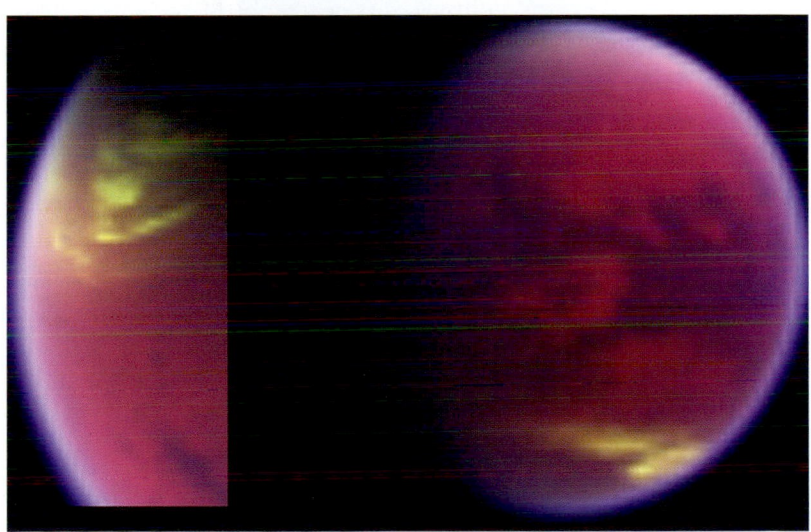

FIGURE 7.13 Clouds on Titan (the colors have been enhanced to bring out cloud features). On the left, the north polar methane clouds are already dissipating as summer approached in May, 2008. The right image obtained in December, 2009 shows an example of clouds in Titan's southern temperate zones. NASA/JPL-Caltech/University of Arizona/University of Nantes/University of Paris. Image processing by Sebastien Rodriguez.

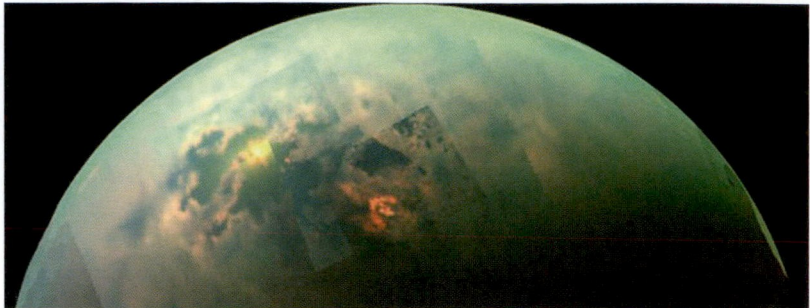

FIGURE 7.15 A *Cassini* visual infrared mapping spectrometer (VIMS) observation of the North Pole of Titan showing that the lakes (the darkest areas) are rimmed by what appears to be a deposit caused by evaporation of the lakes (the bright spot to the northwest of center is another specular reflection; the bright pinkish feature in the center is a cloud). NASA/JPL-Caltech/University of Arizona.

FIGURE 7.18 Frank Paul's "Golden City on Titan" from the back cover of *Amazing Stories*, November, 1941.

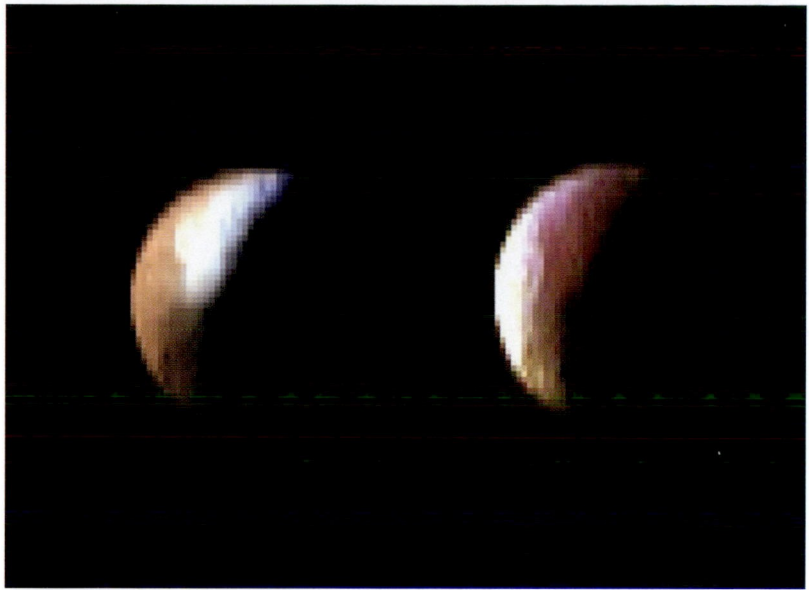

FIGURE 8.7 Two infrared images of Iapetus, one at one micron (left) and the other at 3.5 microns (right). On the left, the polar cap appears bright and the dark side is dark, but on the right the polar caps become dark and the dark material bright. In both wavelengths the light is reflected light and not thermal radiation. NASA/JPL-Caltech. Image processing by Thomas Momary.

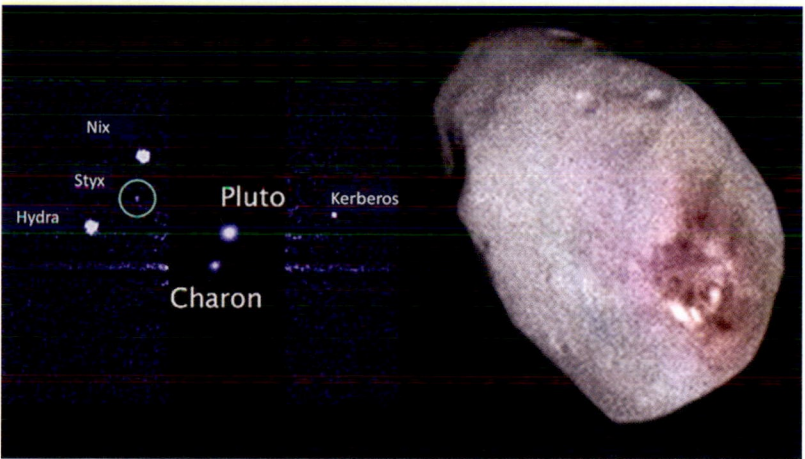

FIGURE 9.3 (Left) Pluto and its moons imaged with the *Hubble Space Telescope*. (Right) An image of Nix returned by the *New Horizons* spacecraft. The longest dimensions of the moons, determined by *New Horizons*, are – Charon: 737 miles; Styx: 4 miles; Nix: 30 miles; Kerberos: 7 miles; Hydra: 34 miles. Pluto has a diameter of 1,448 miles. Like Hyperion, the four smallest moons are in chaotic rotation. NASA/Space Telescope Science Institute/Johns Hopkins University Applied Physics Laboratory.

FIGURE 10.1 The *Hubble* Ultra Deep Field survey, constructed from images obtained in 2003 and 2004, and showing about 10,000 galaxies. NASA/ESA.

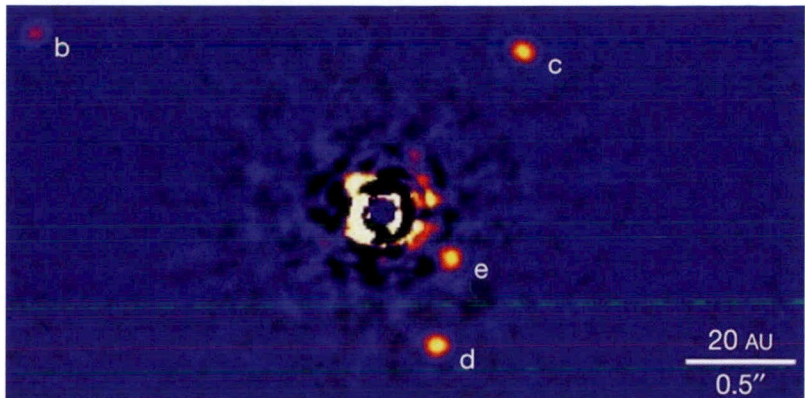

FIGURE 10.3 A direct image of the four planets of HR 8799 obtained on the Keck telescope. 0.5 inches is half an arcsecond, where there are 3,600 arcseconds in one degree of sky. Image by Christian Marois of the National Research Council Canada, Herzberg Institute of Astrophysics and his colleagues. Courtesy University of California.

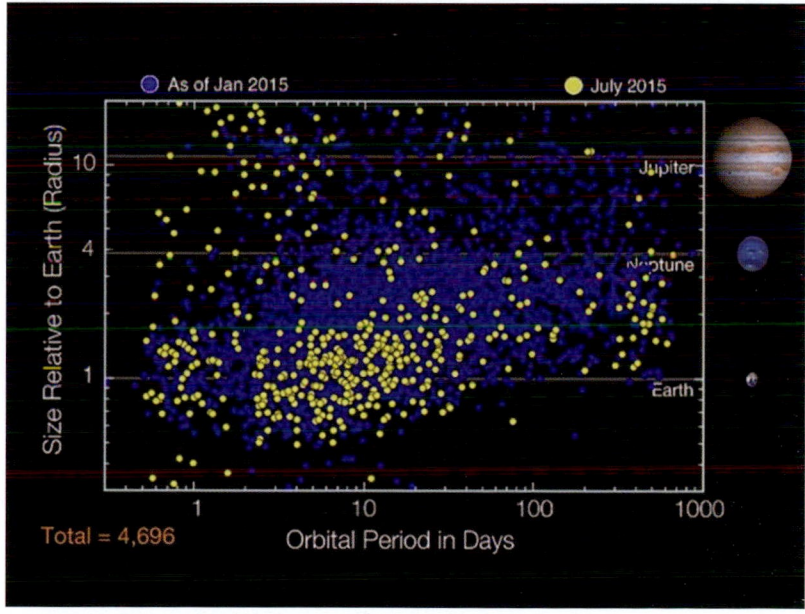

FIGURE 10.4 A summary of the exoplanet candidates detected by *Kepler* as of July, 2015; nearly half have been confirmed. NASA. Graph by W. Stenzel.

FIGURE 10.5 *Hubble* image of the "Pillars of Creation," a star-forming region in the Eagle Nebula about 7,000 light years from the Earth. NASA/Space Telescope Science Institute.

FIGURE 6.1 Trina Ray of JPL.

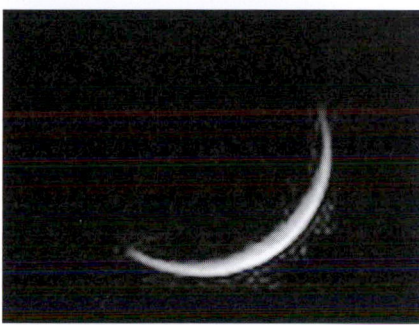

FIGURE 6.2 An enhanced image of Enceladus obtained by the *Cassini* camera on January 16, 2005 when it was 100,000 miles from *Cassini*. A possible plume-like structure is visible off the lit crescent of the moon. NASA/JPL-Caltech. Image processing by Tom Momary.

members of the imaging team to clue them in on our discovery. There was a disagreement on the imaging team at that point. Some members were certain that they had discovered activity on the tiny moon, while another faction led by Carolyn Porco, the *Cassini* Imaging Team leader, was adamant that the features their instrument had detected were not plumes.

Trina, Brian, and other *Cassini* project personnel spoke to Project Manager Bob Mitchell about their findings, and Mitchell spoke to Porco about them. She said in an email to Mitchell that she was certain the tenuous distended feature that was seen in the November 2004 image was not an active plume from the surface of Enceladus. Porco was following the usual conservative approach of the careful scientist to explain an extraordinary scientific phenomenon by a

FIGURE 6.3 An image from the 60-inch telescope on Palomar Mountain during ring plane crossing in 1995. Enceladus is the bright point just to the right of the rings. From left, the other moons are Titan, Rhea, Mimas, Tethys, and Dione. This is an enhanced version of an image obtained on the same observing run as the image shown in Figure 7.1, stretched to show the inner moons.

pedestrian explanation, similar to the tack taken at the time of Linda Moribito's discovery of Io's plume 25 years earlier (see Chapter 5). But some members of Porco's team, as well as our team at JPL, had the hunch that something was there. Events would soon unfold that show again how disagreement drives scientific discovery.

Enceladus had already acquired a reputation as a mysterious and intriguing object. The moon is usually swallowed up by bright light from the rings and body of Saturn, but once every 15 years or so, when the rings appear edge on to the Earth and seem to disappear, the tiny moon pops up from the black sea of space surrounding Saturn. Figure 6.3 is an enhanced version of an image I took while I was observing with my colleagues Dick French, Phil Nicholson, and Colleen McGhee at the 60-inch telescope on Palomar Mountain during such a "ring plane crossing" in 1995. The rings appear as tiny "ears" (to use Galileo's expression), but the inner satellites are bright and clear. Enceladus was discovered by William Herschel on August 28, 1789, during another Saturn ring plane crossing. Earlier that year, Herschel had finished his largest telescope, the clunky 40-foot that reigned as the most massive telescope for over 50 years. In 1795, he wrote that "having brought the telescope to the parallel of Saturn, I discovered a sixth satellite of that planet [the others were Tethys, Dione, Rhea, Titan, and Iapetus, all discovered between 1655 and

1684 by Huygens or Cassini]; and also saw the spots upon Saturn, better than I had ever seen before, so that I may date the finishing of the 40-foot telescope from that time."[1] Three weeks later Herschel discovered the seventh moon of Saturn, Mimas, but the more agile 20-foot instrument remained Herschel's favorite. For Herschel and his assistants, the changing of the 40-foot behemoth's one-ton mirrors caused "many hair-breadth escapes from being crushed."[2]

My viewing of Enceladus in 1995 was the only time I ever saw this moon from an Earth-based telescope. (During the next ring plane crossing in 2009 I was too inundated with data from *Cassini* to work on any telescopic data.) I had been studying the moon for nearly 15 years from *Voyager* and *Cassini* images, but I had never seen the little moon in real time (well, almost real time – it takes light over an hour to reach us from Saturn's distance. But when data are gathered from spacecraft it takes about a day for a scientist to receive it because it has to be stored on the flight computer first until there are enough data to relay to one of JPL's large radio telescopes that are part of the Deep Space Network). Fourteen years later, while I was at the Royal Observatory Greenwich in London for the June 2009 *Cassini* Project Science Group meeting and doing an interview for *National Geographic*, I saw the single remaining fragment of Herschel's 40-foot instrument, lovingly posed in the Observatory gardens. Figure 6.4 is a picture I captured of this sublime moment, and Figure 6.5 shows the picture John Herschel (son of William) took of the same telescope 169 years earlier, at the dawn of photography, during the instrument's decommission. Note that the tube itself is on its side; apparently John was afraid the rotten tube would crumble and hurt his children.

When the two *Voyager* spacecraft hurtled by the moon in 1980 and 1981 to transform Enceladus from the tiny light seen in Figure 6.3 to a tangible world, they found a strange body indeed (see Figure 6.6). Part of the moon was covered with the scars of impact craters and was at first glance old looking – your typical icy satellite. But other parts of Enceladus appeared to have undergone melting and refreezing, as if the moon had been subjected to a catastrophic period of high activity.

FIGURE 6.4 A small segment of Herschel's 40-foot telescope at the Royal Observatory Greenwich.

FIGURE 6.5 John Herschel's photograph of the same telescope, obtained in 1849, near the dawn of the age of photography.

These parts of the surface were smooth and crater free (see Figure 6.6). But Enceladus was unlike any other moon that showed activity: both old and young parts of the moon were very bright and reflective. In fact, they were as bright as freshly fallen snow on Earth. Other moons in the Solar System – for example Ganymede from the last chapter – have old and young areas, but in general the older areas are darker, mainly from accreting interplanetary dust – the same dust that falls on the Earth and sometimes makes it to your coffee table (although

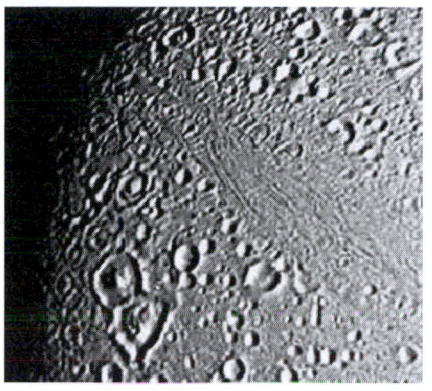

FIGURE 6.6 The best *Voyager 2* image of Enceladus, taken about 54,000 miles away on August 26, 1981. Most of the moon's terrain looks old and cratered, but there is a smooth plain in the center, suggesting some active geological process on the moon. Even more intriguing, the surface reflects nearly 100% of the sunlight falling on it, as if it were covered in freshly fallen snow. NASA/JPL-Caltech.

as mentioned in Chapter 4 *most* of the dust in your living room is from more mundane sources such as lint, cat dander, and just plain dirt). Scientists can even estimate how old a surface is by measuring how dark it is. But Enceladus was not dark *anywhere*. The young, resurfaced areas were bright, but the older cratered regions were just as bright. The moon just looked like its entire surface was covered in bright snow.

I often wonder what it would be like to just walk the surface of this sparkling winter wonderland, or to ski on it. And to planetary scientists, bright signals fresh and new, something that could only be produced by an active geologic process. Moreover, this process would have to propel material above the surface and subsequently cause it to "rain" down upon the entire icy world. A geyser or a volcano with a large plume seemed to be the only thing that could possibly explain the appearance of Enceladus. But after all the ridicule Al Cook and his team had received over their claim of an active geyser on Europa, putting a plume on Enceladus was a hard sell.

Kevin Pang was the first planetary scientist to claim that Enceladus was erupting ice volcanoes. My colleague Anne Verbiscer of the University of Virginia made a very definitive claim for geysers on Enceladus, and reiterated an idea from an earlier paper I wrote with Joel Mosher and Torrence Johnson that bright ice particles from geysers were being launched into space and coating the other moons of

Saturn ("Cosmic graffiti artist caught in the act," as her *Nature* paper stated).

Another mysterious aspect of Enceladus is that it is bathed in the tenuous, bloated E-ring of Saturn. This ring is even thinner than air on the surface of the Earth. We are all familiar with the bright, charismatic main ring system of Saturn. It is one of those breathtaking sights you remember the first time you see it through a telescope, or when you view the spectacular images sent back by the two *Voyager* spacecraft of fine ringlets, seemingly etched on the sky by the stylus of a giant cosmic record player. It is what makes Saturn famous. As the *New Yorker* stated so succinctly in a poem years ago: "Why study Saturn? Because of the rings!" But the E-ring is so faint it wasn't even discovered until 1980, when Bill Baum and his colleagues at Lowell Observatory placed a super-sensitive new-fangled – then, at least – charge-coupled device (CCD) being developed as a training camera for the *Hubble Space Telescope* Science Team on a 60-inch telescope owned by the US Naval Observatory. This telescope was not among the largest in the world at all. It was the new sensitive detection powers of the CCD that led to the detection of the E-ring, illustrating that the introduction of a new technology often leads immediately to new discoveries. Now that technology is part of every cell phone camera. Baum and his colleagues realized that the ring was of recent origin – otherwise it would have flattened out into a much thinner ring. Later scientists noticed that the densest part of the E-ring is at the orbit of Enceladus, as if the moon were the source of the ring.

Was the E-ring named after Enceladus? No. Scientists weren't aware of any connection between Enceladus and the E-ring when it was first discovered: Saturn's main rings are named sequentially after the letters of the alphabet: A, B, C, D, and so on. E just happened to be the next letter in line. But as we shall see, it was a fitting designation.

There had also been nagging observations of Enceladus by generations of astronomers who hinted at its weirdness. Percival Lowell makes another appearance in 1913, when he and Lowell Observatory observer E. C. Slipher reported that one side of Enceladus was about

30% brighter than the other. Then in 1972 to 1973, Otto Franz and Robert Millis, also of Lowell Observatory, observed a similar brightening on the same side. Moreover, these brightenings both occurred when the south pole of Enceladus was visible from Earth. I remember sitting for hours in my graduate student's office perplexed by the observations of Otto Franz and Bob Millis. Their observations did not agree with what I was seeing in the *Voyager* data – a uniformly bright world – but I knew that Franz and Millis were very careful, reliable observers, so it was extremely unlikely that their data were bad. It is often these nagging little details – things that just don't quite fit together – that mean a discovery is on its way.

After the reconnaissance mission of the two *Voyagers*, which visited all four of the giant gaseous outer planets, Jupiter, Saturn, Uranus, and Neptune, NASA embarked on a program of sending individual spacecraft for closer scrutiny of each one. *Galileo* studied Jupiter and its moons for nearly eight years, starting in 1995. The *Cassini* spacecraft was built to launch in 1995 for a four-year study of Saturn, its moons, and rings, to start in 2002 (the mission was extended for three additional periods: it is now scheduled to end in 2017). Named after Giovanni Domenico Cassini (1625–1712), the Italian astronomer and astrologer (yes, the two disciplines were not yet separate) who was the first director of the Paris Observatory, the program was originally to be part of a two-for-one program. *Cassini* was going to Saturn, but a second spacecraft, the *Comet Rendezvous Asteroid Flyby (CRAF)* was to greet comet Kopff for a three-year tour and the bonus of a flyby of an asteroid on the way. *CRAF* was the first mission to be developed, but when budgetary competition between the two missions developed, *CRAF* was canceled and all its funds were redirected to *Cassini*. I took the cancellation of *CRAF* pretty hard, because early in my career I had been selected by NASA Headquarters to serve on its elite Imaging Team: all that work for nothing. But I was having a lot of fun on *Cassini*, which was delayed for two years after further budgetary and technical problems. This Cadillac among spacecraft has 12 instruments, including four cameras to detect

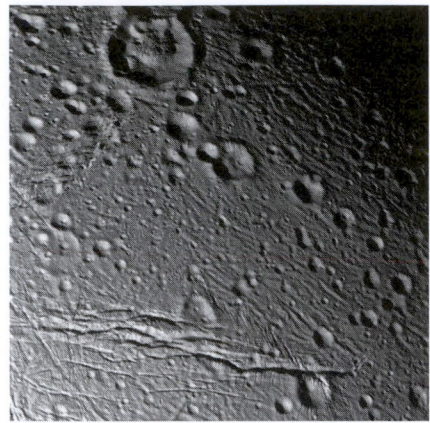

FIGURE 6.7 An image from first targeted flyby of Enceladus by *Cassini* on February 17, 2005, which approached to within 700 miles. NASA/JPL-Caltech.

signals all the way from the ultraviolet to the infrared, a radar mapper, and several instruments to detect particles and magnetic fields in the area around Saturn. The mission was executed as a joint program with the European Space Agency, with the Europeans building the sophisticated *Huygens* probe to land on Titan, the giant moon of Saturn (see Chapter 7).

Our excitement ran high as *Cassini* made its first close pass by Enceladus on February 17, 2005, approaching to within 700 miles. The spacecraft captured the illuminated face of Enceladus crossed by giant cracks and faults, and fresh, bright, snow-like deposits coating the entire moon, but there was no hint as to how this wintry landscape formed (Figure 6.7). During the next flyby, on March 9, *Cassini* swooped to within 450 miles of the surface of Enceladus, to provide more similar views of the ice world: intriguing but no smoking gun. This feat of accurately bringing the spacecraft so close to the surface of the moon while their relative velocity was about four miles per second was perfectly executed by the legendary Navigation Team at JPL. Their skills are akin to throwing a basketball from the Rose Bowl in Pasadena, California and landing it in the hoop at Madison Square Garden in New York City.

But now Enceladus was beginning to reveal some of its secrets. *Cassini*'s magnetometer detected a draping of the magnetic field of

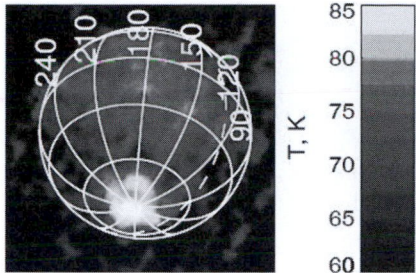

FIGURE 6.8 John Spencer's infrared image showing a large heat anomaly in the southern hemisphere of Enceladus. Courtesy John Spencer. See plate section for color version.

Saturn around the southern tip of Enceladus, as if it had met a barrier to its smooth flow. Previously the Cosmic Dust Analyzer had found evidence for a thin atmosphere. Michele Dougherty of Imperial College London, the Principal Investigator of the Magnetometer, made a pitch to the Project to lower the closest approach of the next flyby on July 14 to a breathtaking 102 miles. Most other *Cassini* scientists supported her request, and Bob Mitchell approved it, after conferring with his spacecraft safety team. One of the last holdouts continued to be Carolyn Porco, who sent an email on April 14 to John Spencer of the Southwest Research Institute, a member of the Composite Infrared Spectrometer (CIRS) Team, and several other *Cassini* scientists that stated "This sounds crazy to me ... for the Project to tweak the tour when we're not sure that the 'thing' is real. I have encouraged my team members to act on this quickly to verify whether or not it is real. The last I heard is that it wasn't."

A few days after the July 2005 flyby, Spencer sent a few *Cassini* scientists a wondrous and historic image of Enceladus (Figure 6.8). We have all seen those infrared images of tigers burning bright in the forests of the night, or of lost people appearing in the cameras of heat-seeking helicopters. Well, Spencer showed us an infrared image of the south pole of Enceladus – the part of the moon that should have been the coldest, and devoid of a significant infrared signal. It was a blaze of heat! He had discovered the "smoking gun" that we were all seeking. The first view of the Pacific Ocean by Lewis and Clark, the first view of Half Dome by John Muir, the first view of Monument Valley

FIGURE 6.9 An image of the plume obtained on November 30, 2010 seemingly composed of smaller jets. JPL PIA 17184.

by Everett Ruess, and your own first view of the Grand Canyon – the thoughts and feelings evoked by these sightings were ours as we viewed that image. *Voyager 2* had never seen the South Pole of Enceladus because of a technical glitch: the scan platform that moved the camera and could have pointed to the South Pole had become stuck. Finally, in November 2005, Porco and her team, who now fully accepted the idea of geysers on Enceladus, used the *Cassini* camera to shoot a picture of a glorious crown of plumes perched on the globe of the moon. Even better images were soon gathered such as the one in Figure 6.9.

The area of the greatest heat was the location of one the oddest terrains on any planet or moon – a crater-free area filled with tiny grooves and four bigger cracks about 300 m (984 ft) deep called "tiger stripes" (Figure 6.10). In the whimsical way that the International Astronomical Union names features on the planets, the four tiger stripes are eponyms of cities in the Middle East: Cairo, Baghdad, Damascus, and Alexandria. Boulders and large rocks surround the tiger stripes, as if they were ejected during particularly violent eruptions (Figure 6.11). A bluish hue surrounds the stripes, perhaps

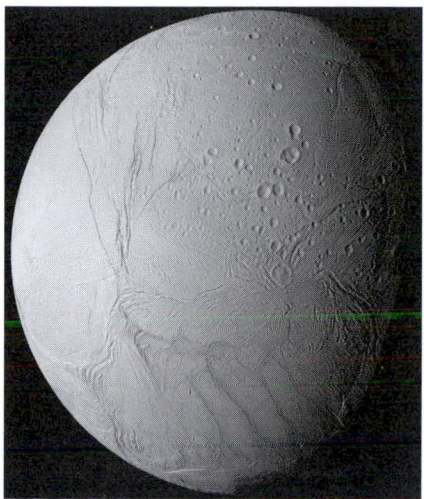

FIGURE 6.10 The tiger stripes in the south pole of Enceladus are named after cities in the Middle East: (left to right) Damascus, Baghdad, Cairo, and Alexandria. NASA/JPL-Caltech. See plate section for color version.

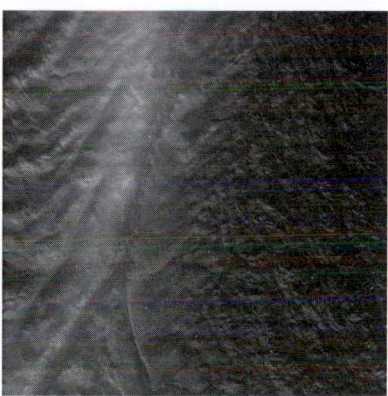

FIGURE 6.11 The flyby of August 13, 2010 showing the tiger stripe Damascus caught between day and night. Icy particles appear to be arising from the surface openings. NASA/JPL-Caltech.

because the particles that stay near the vents are heavier and bigger. One image taken at dusk shows what appear to be icy particles rising from a vent (Figure 6.12). Molecules containing the building blocks of life – hydrogen, carbon, oxygen, and nitrogen – were discovered in the plumes by *Cassini*'s particle detectors and on the surface by the Visual Infrared Mapping Spectrometer (VIMS). These molecules are often called "prebiotic" or "organic," because they are necessary for life to arise and to be sustained.

The energy production from the activity on Enceladus is enormous. Recent work by John Spencer and Carly Howett, also at

FIGURE 6.12 The flyby of November 21, 2009 showing boulders and the tiger stripes close up. The image is about seven miles on each side. NASA/JPL-Caltech.

Southwest Research Institute, suggests the heat output is about 5 GW – 5 followed by 9 zeros, in watts, the scientific unit for power. This is enough power to heat 4 million ovens running at 350°F. Another way of understanding this energy: Enceladus provides enough constant energy to power a city the size of Philadelphia.

Why is this tiny world the only icy moon that has a boiling cauldron? We don't know yet. We are still at the stage of gathering all the data we can and producing models that make sense. But most models of Enceladus's activity center on two main sources of energy: radioactivity and tidal effects. Of the icy moons of Saturn, Enceladus has the largest fraction of rocky material in its core (except for Phoebe, which was probably captured as it wandered in from the outer reaches of the Solar System). Radioactive elements such as uranium and thorium tend to be chemically bound with rocks, so the more rock a moon has, the more heat from radioactive elements it will produce. Such radioactive decay is the source of most internal heat on the Earth, and the driver of terrestrial geologic processes. But for Enceladus, there isn't enough radioactive heat to drive the plumes. Tidal energy from Saturn must add some heat. Enceladus is also in a special dynamical relationship with Dione: Enceladus orbits twice for each time Dione orbits. This orbital mean motion resonance causes an extra "tug" by Dione each time they are near each other in their orbits. The tug isn't as great as that acting on volcanically active Io by Europa and Ganymede (Chapter 5), and most scientists don't believe the tidal forces are sufficient to explain the vast amount of heat emanating from the south pole of Enceladus. Francis Nimmo of the University of California at Santa Cruz suggested that frictional heating from the motion of faults below the tiger stripes may provide additional heat. Perhaps the activity is cyclic, where heating, cracking, cooling, and closing of the vents occurs in spurts over tens of millions of years. If this idea is correct, we are just lucky enough to see Enceladus in one of its active periods. "Fossilized" tiger stripes near the equatorial regions of Enceladus speak of past episodes of activity on the moon.

An important point that was unresolved until 2014 is whether liquid water is powering the jets. Almost from the time we discovered the plume on Enceladus, most scientists believed a liquid ocean underneath the surface of Enceladus provided the material for the plume, and that as the water is expelled under high pressure it is rapidly quenched into small spherical particles. This transition from liquid to solid produces the heat seen at the surface. Susan Kieffer of the University of Illinois presented a counter-theory that the activity is powered by chemical reactions in ammonia that is chemically combined with water ("clathrates" in the chemist's lingo). One piece of evidence strongly implies Enceladus has a liquid ocean: the particles in the plume are frozen salt water. On the Earth, salt water in the oceans is formed by the endless cycle of rain dissolving, washing out, and transporting minerals such as salt to the Earth's seas. On Enceladus, the only way to get salt into the plume particles would be for a submerged ocean to dissolve it out from a rocky mantle in contact with liquid water. Finally, Candice Hansen of the Planetary Sciences Institute and the Ultraviolet Imaging Spectrometer team showed that particles in the plume are supersonic, which strongly suggests they are being vented as liquid jets.

Even more proof arrived for an ocean under the surface of Enceladus that was powering the jets. My colleague Jay Goguen performed a very clever experiment with VIMS during a flyby on April 14, 2012 when *Cassini* approached to within 50 miles of the surface. He designed an observation to put the instrument in "point" mode and to drag the detector across the face of the tiger stripes. Goguen and his team (which included me) were able to show that the emitting regions for the plumes were hot and small – about 200 kelvin (–100°F or –73°C), still cold by terrestrial standards, but 250°F (121°C) warmer than it should be – and only about 30 feet in size. Then another VIMS Team member, Matt Hedman of the University of Idaho, showed that the plume was more intense when Enceladus was further from Saturn, when tidal forces were smaller and the cracks at the tiger stripes would open up to let vapor escape. The Radio Science Team analyzed how the tiny moon pulled the spacecraft during its approaches, and the best conclusion was that there was a liquid ocean lying beneath the tiger stripes that extended about halfway up to the equator. Finally, Peter Thomas of Cornell University and his colleagues analyzed wobbles in the orbit of Enceladus and concluded that the ocean had to be global, similar to Europa's alien sea.

Meanwhile, Porco and her team were busy locating the position of over 100 jets seen in *Cassini* images. She confirmed that they came from small hot spots that Spencer and Goguen and their teams had mapped on the tiger stripes. Frictional heat could not be a major source of heat because the entire tiger stripe was not very hot and yet it was ejecting a great mass of water vapor. It is important to recognize that the models of Sue Kieffer on ammonia clathrates and Francis Nimmo on frictional heat were completely valid theories at the time they were presented. Conflicting theories compel scientists to gather more data and arrive at scientific truth. Without disagreement science does not progress. Nimmo in fact did some later theoretical work on the nature of the tides in the liquid ocean and its interaction with the surface of Enceladus. And later Joseph Spitale of the Planetary Science Institute presented convincing evidence that many of the jets were not real. Instead, the water was ejected as a curtain, and that many of the jets were just more dense regions of the curtain, or sections seen edge-on.

Enceladus was literally "bursting at the seams" (as JPL described in a press release), but why are the seams only at the south pole? The tiger stripes are in a basin, but why is the basin where it is? Bob Pappalardo and Francis Nimmo developed a clever idea from classical geology that a diapir, a lighter blob of material, moved up from the interior of Enceladus into the more brittle ice crust of the moon. This process is similar to the wax rising in a lava lamp. Once the material rose to leave a density deficit in the interior of the moon, the rotation axis realigned to the most stable configuration, which was to have it where the density was lowest.

One question we haven't answered yet is how variable the geysers on Enceladus are. For all of you used to sitting on that bench at Yellowstone and waiting for Old Faithful to come out at its appointed times, we haven't seen anything like that on Enceladus. So far the plumes appear to be fairly stable, over time periods of years.

Enceladus is truly a world fantastic, but in one way it is a world familiar: it has a habitable environment. When we explored Europa, we found that it almost certainly has a liquid ocean below the surface. Just as Europa does, Enceladus possesses the three main requirements for life: liquid water, food (the "prebiotic" molecules), and a source of energy. In 2015, NASA announced that there are warm jets of water erupting on the floor of the subsurface ocean: these "smokers" are ideal locations for life to form, similar to the oceanic vents around which life on Earth is believed to have arisen. There is currently no evidence that life exists on Enceladus and, if it did, it would probably be bacteria or other primitive life forms. NASA is currently studying a mission to Enceladus that would include a radar sounder to peer below the surface and to map the extent of any ocean. Seismometers on the surface would reveal the depth of the ocean and its extent in great detail. But such a mission is very expensive and there are currently no funds in NASA's budget to start this attractive mission.

It took us a long time to discover the plumes and the hot spot on Enceladus. Both *Voyager* spacecraft flew by Enceladus without discovering any of the good stuff on Enceladus, and during the first close targeted flyby by *Cassini*, the moon's active region never entered the spacecraft's field of view. Is it possible that other moons of Saturn have plumes or other types of activity waiting to be discovered? In 2007, Jim Burch of the Southwest Research Institute and his colleagues discovered that both Dione and Tethys seemed to press against the

FIGURE 6.13 Dione shows possible cryovolanoes in the middle, with some cracks in the surface and global powdery deposits. NASA image processed by Paul Schenk.

magnetic field of Saturn. They attributed this interaction to streams of particles coming from the moons. Krishan Khurana of the University of California at Los Angeles and a member of the magnetometer team reported an outflow of material from Dione of about 0.6% (less than a hundredth) of the material from Enceladus. Perhaps even more intriguing, an image of the south pole of Dione shows what appear to be old, perhaps inactive, tiger stripes, and Paul Schenk of the Lunar and Planetary Institute has identified what he believes to be ice volcanoes on Dione (see Figure 6.13). These volcanoes are surrounded by what appears to be a huge plain of fine particles. Could this plain have been created by material expelled from the ice volcanoes and reaccreted back on to the surface? Roger Clark at the Planetary Sciences Institute detected a transient atmosphere around Dione in 2005 with one of the *Cassini* instruments, but he hasn't seen it since. In spite of vigorous plume searches, we haven't discovered any plumes yet from Tethys or Dione. Could the "plumes" causing the disturbances in Saturn's magnetic field just be anemic puffs outgassing from an old, almost inactive tiger stripe or a small vent? We just don't know. We will continue to look, but the question of activity on Dione may never be answered in the lifetime of the *Cassini* mission.

Speaking of lifetimes, all spacecraft eventually die, like creatures of flesh and blood. NASA does not like to fund missions forever,

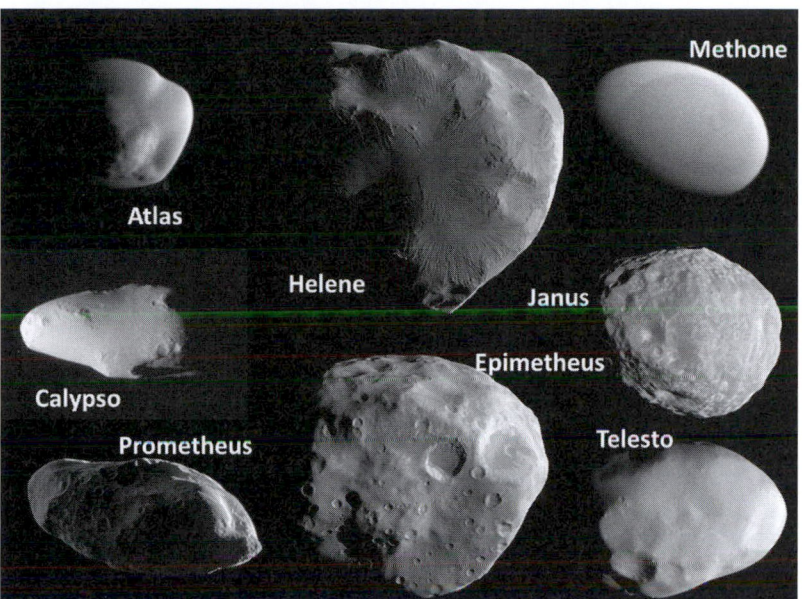

FIGURE 6.14 A montage of some of the small moons of Saturn, showing diverse surfaces ranging from ones that are heavily cratered to those that are entombed in a coat of icy particles. These particles have tumbled down almost like avalanches from the rugged crags of Helene and formed a little skirt around Atlas. The sizes of the moons are not to scale. Their mean diameters in miles are: Atlas, 19; Helene, 22; Methone, 2; Calypso, 13; Prometheus, 53; Epimetheus, 72; Janus, 110; and Telesto, 16. Figure constructed by the author, based on NASA images.

as new and better projects are waiting patiently in line to come to life. We *Cassini* scientists have decided to end our spacecraft violently, by plunging this marvel of human engineering into Saturn. At the end of the mission in late 2017, *Cassini* will orbit Saturn in an ever-tightening series of orbits, including some daring passes between the rings and the planet itself. This death spiral will offer unprecedented views of the great globe of Saturn and its rings, with a breathtaking final descent into the depths of the planet. We will swoop to within a few thousand miles of some of Saturn's strange small moons (jokingly called "Rocks" by our Science Team). Figure 6.14 shows a montage of these strange moons compiled for our NASA "Senior

Review" in 2014 when a team led by *Cassini* Project Scientist Linda Spilker argued before NASA Headquarters that we should be allowed to explore until 2017.

In *Cassini*'s final plunge into the cloud tops of Saturn, the pressure of the planet's thick atmosphere will crush the spaceship and eventually dissolve its parts to mix with the stuff of Saturn: a fitting grave to this gallant robot. We will be taking data as long as we can. With that first touch of humanity to the farthest planet known to the ancients, Enceladus, which peeked out behind the veil of scattered light from Saturn to reveal its face, will go back into hiding, awaiting another generation of scientists to step forward and lift its veil again.

NOTES

1 Herschel, W. 1795. Description of a forty-feet reflecting telescope. *Philosophical Transactions of the Royal Society of London* 85, 347–409.

2 Herschel, C. 1879. Memoir and correspondence of Caroline Herschel. Edited by Mrs. John Herschel, quoted in *The Age of Wonder*, by Richard Holmes (Pantheon, New York) 2008, p. 191.

# 7   Titan: An Earth in Deep Freeze?

My first glimpse of Titan in real time wasn't through a toy telescope or even through the respectable telescopes that one finds at star parties. It was a hot August in 1995 and I was at Palomar Observatory during Saturn ring plane crossing, that rare time when the bright rings appear edge-on and nearly disappear to reveal the whole of Saturn's six major inner moons, encircling the planet like a string of pearls, suspended on some imaginary scaffolding in the skies. The biggest gem of all was Titan, a ghostly white orb silently looming in space, lit with the soft reflected light of the Sun, inviting us to come closer (Figure 7.1). I had seen much better images from *Voyager 1* depicting this mysterious, cloud-enshrouded world, but never had the moon been so palpably close as during this live-action view. I could not have imagined the enchanted world that lurked beneath those clouds.

Titan was discovered in 1655 by Christiaan Huygens (1629–1695), a Dutch astronomer and mathematician who designed and constructed technologically advanced telescopes, which again shows how technology drives science. He is perhaps best known for his discovery that the velocity of light is finite. Huygens even built a refracting telescope with its lenses in the open air – no unwieldy tubes to drag it down (Figure 7.2). But Huygens's clever invention wasn't very practical: it was just too difficult to align, and there was no way of keeping out stray ambient light. Huygens's great discoveries were made with more pedestrian "bread and butter" telescopes. Titan, for example, was discovered with a 12-foot, 50 power telescope that is today surpassed by the typical equipment in a college's observatory.

Most moons of the Solar System appear as fading tenuous blinks of light in the sky, even when seen through a moderately sized telescope. Intently studying these tiny points, astronomers have used

FIGURE 7.1 The moons of Saturn during the ring plane crossing in August, 1995. Titan is to the far left. Image by the author with the 60-inch telescope on Palomar Mountain. Other observers with me were Phil Nicholson of Cornell University, and Richard French and Colleen McGhee of Wellesley College.

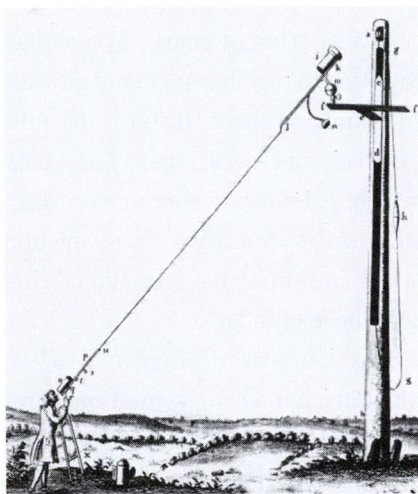

FIGURE 7.2 Huygens's odd telescope, 123-feet long, without a barrel. Completed in 1686, it was meant to be light and portable, but it was exceedingly difficult to align, and Huygen's didn't make any of his great discoveries with it.

their clever tricks through the ages to infer a surprising amount of information. For example, if a rotating moon or planet dims and brightens in a regular fashion, one can be sure – at least if the body is round – that one hemisphere contains materials that are much brighter than the other – perhaps patches of frost. By studying the color of surfaces and atmospheres, astronomers can deduce their composition. For example, water ice, the main component of many moons that orbit the gassy outer planets (Jupiter, Saturn, Uranus, and Neptune), is blue-colored, while sulfur, the main component of the Galilean satellite Io, is reddish.

The great Dutch–American astronomer Gerard Kuiper (1905–1973) – sometimes called the father of planetary astronomy, and the founder of the Lunar and Planetary Laboratory at the University of Arizona – used one of these seemingly magical techniques of astronomy to make the unexpected discovery of a thick atmosphere on Titan. Observing Titan in 1943 to 1944 with the McDonald Observatory's 82-inch telescope, with which the moon appears about twice as bright as it does in the Palomar image shown in Figure 7.1, he detected the tell-tale signature of gaseous methane. He correctly understood from the intensity of the signature that the atmosphere was thick, the only such atmosphere on any moon. In a technique called spectroscopy, Kuiper measured the brightness of Titan as a function of color – its spectrum – and it decreased exactly where methane does. In his 1944 *Astrophysical Journal* paper, he boldly stated that "the total thickness of the atmosphere is comparable to, but somewhat less than, that of the observable layers of Saturn and Jupiter." What an honor for an object just a tiny bit larger than Mercury! There are many statements in the popular literature that say a spacecraft discovered Titan's thick atmosphere (e.g. the Wikipedia entry for *Voyager 1* says "*Pioneer 11* had one year earlier [1978] detected a thick, gaseous atmosphere over Titan"). Not at all – the thick atmosphere was discovered a generation earlier with "bread and butter" astronomical techniques, when Titan was still just a dot. Kuiper also noted Titan's orange hue, which we will see is a clue to its importance as a sort of astrophysical laboratory.

Another clever observation of this barely resolved object came when Joseph Veverka of Cornell University – my advisor in graduate school – observed Titan in the late 1960s with the 61-inch telescope at Harvard College Observatory. Veverka was observing with a polarizing filter, similar to the polarizers on dark glasses. Polarized light just means that the orientation of the waves that make up the light are all lined up. He noticed that this "lining up" of light coming from Titan was more like that of bodies with atmospheres rather than bodies with bare surfaces. Veverka deduced that Titan must have a thick cloud deck with a thinner atmosphere above it. This view was validated in 1980 when the *Voyager 1* spacecraft made a close flyby to the moon to reveal Titan's atmosphere and its detached haze layer, a mass of sooty, smoggy particles that rival anything in Los Angeles. The atmosphere continues to fade out into space so slowly that our early flybys executed by *Cassini* could go no lower than 600 miles; otherwise the haze and molecules in the atmosphere might heat up the spacecraft or damage parts as we whizzed by at 10,000 miles per hour.

Earth's gravity hugs our atmosphere to the planet so it is much more "compressed" than that of Titan. The most recent measurements give a surface pressure to Titan's atmosphere of 1.45 times that of Earth: this is the most Earthlike pressure in our Solar System, and the only substantial atmosphere on any moon. The composition is also Earthlike: mainly nitrogen with methane, the second most common gas, instead of Earth's oxygen. Kuiper didn't discover nitrogen because it doesn't have the obvious dips in color that methane does: we had to wait for *Voyager 1* for that discovery.

Imagine being an astronomer painstakingly studying a tiny illuminated point for years, when in the space of a few months or days, the resolution increases by a factor of 10, and then 100, and then 1,000, and finally 10,000. Saturn and its moons are almost a billion miles from Earth, but the first spacecraft to visit the system (*Pioneer 11*) came more than 100 times closer to Titan. To illustrate what that means with an image of something that is familiar, look at the photograph of two people standing across the mall at JPL at a distance of 300 feet (Figure 7.3). They are barely a dot, indistinguishable from the clumps of distant trees. But bring these two people 100 times closer, to three feet, and you can see them clearly. *Voyager 1* brought Titan another factor of 100 closer – the equivalent of seeing the people in the photograph from an inch away. In addition to new views brought by increased clarity, advances are always made when we look at additional wavelengths. *Cassini* not only came ten times closer than *Voyager 1*, it possessed a collection of cloud-piercing instruments: radar and infrared cameras. The history of our exploration of Titan is a series of peeling away successive layers of mystery to reveal a world

FIGURE 7.3 Kim Tryka and Michael Hicks are standing on the far side of the large lawn in JPL's mall areas. They are just dots (if you know where to look). Bringing them 100 times closer transforms them from dots into real people, analogous to the way spacecraft transform a dot in the sky into a geologic world. Photos by the author. See plate section for color version.

that is more like the Earth than perhaps any other planet or moon. New technologies have opened our eyes to perceive a world that is a sister planet in many ways. Some astronomers have referred to Titan as an Earth in deep freeze.

*Voyager 1* was the first mission to come in close to Titan, and it was also the first mission in which I participated as a "real scientist." So important was it for scientists to study enigmatic Titan that they decided to make a close pass of less than 15,000 miles to the moon. But Titan flung *Voyager 1* out of the Solar System, making it impossible for this precious asset to study additional planetary targets; *Voyager 2* went on to study Uranus and Neptune and their moons. *Voyager 1* could also have been redirected to Pluto – which was still a planet back then, but Titan was seen to be more interesting. (As we shall see in Chapter 9, that assumption is doubtful.) We got to Pluto anyway – when it was no longer a planet, but 35 years later, in what might be the *last* mission I will work on. During this first mission in which I was participating in the discovery phase, I got to glimpse surprises while working at the side of Joe Veverka while I was his graduate student at Cornell. I often felt like a baby duck, one of a few graduate students following Joe around at conferences and important events. He was – and still is – a great mentor. It wasn't until I was a fully-fledged

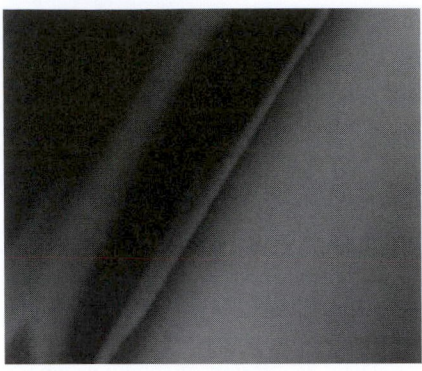

FIGURE 7.4 One of the closest images obtained by *Voyager 1* of the orange atmosphere of Titan and the thick smoggy layer of haze extending out into space. The image has been enhanced in color to show the haze layer and the reddish atmosphere of Titan. NASA/JPL-Caltech. See plate section for color version.

scientist that I viewed new and even more exciting observations of Titan from the inner circle, as we shall see. I wasn't at JPL during the *Voyager* encounters at Saturn – I was too busy writing my PhD thesis at Cornell – but during one of the encounters a local TV station came by Cornell to talk to some scientists about the wonderful discoveries that *Voyager* was making. All the professors were at JPL, so it fell on me to talk about how the *Voyager* spacecraft was rewriting the textbooks. The only person of any note who was left behind at Cornell was Freeman Dyson, the great physicist visiting from Princeton who made profound discoveries in quantum electrodynamics. I sat with him on the panel! As a mere 20-something graduate student, I had never done anything like this before, and I was quite apprehensive. But what I found is that most of the questions people ask are elementary, generally having to do with how big, how far, and how fast. Then of course astronauts: When will we be sending people there? Why weren't any astronauts on board *Voyager*? My usual answer to these questions is that we don't have enough money to send people there during the lifetime of any living person, and that I feel we are really there, peering through the eyes of the instruments on our spacecraft.

But those images of Titan sent back by *Voyager 1* continued to frustrate and tease. Figure 7.4 shows the thick atmosphere and haze layer that was impermeable to visible light, so the moon's surface was still veiled. But ground-based astronomers continued their clever tricks. In the infrared part of the spectrum, where our eyes can't see, there are several "windows," wavelengths (colors) that are clear to sunlight. With infrared detectors, we can "see" through to the surface in these windows. Observing with the giant infrared telescopes sitting high atop Mauna Kea in Hawaii, a team led by Caitlin Griffith, a professor at the University of Arizona, derived the color of Titan's surface within these windows to detect a signature characteristic of water ice. So maybe Titan wasn't so different from the other icy moons that orbit the giant gas planets. Another remarkable observation was made when scientists trained cloud-piercing radar telescopes onto Titan. Don Campbell of Cornell University and later G. J. Black of the University of Virginia and their teams used the giant radio dish in Arecibo, Puerto Rico, which was formed from an enormous sink hole (see Figure 1.4) to watch Titan over the 2000 to 2008 period spanning the early discoveries made by the *Cassini* mission. Black observed radar echos that were characteristic of water ice, with varying amounts of contaminants, but his team also detected narrow specular reflections characteristic of very smooth surfaces, possibly even an ocean.

These glints of light that seemed to sparkle from the surface of the moon could even indicate the presence of a liquid on Titan's surface. Moreover, these reflections became rarer as time wore on. Could there be evaporating lakes or ponds on Titan? Theoretical models by Jonathan Lunine (now at Cornell) and his colleagues suggested substantial amounts of liquid on the surface of Titan, but the liquid couldn't be water. Water is as solid as a rock on Titan, which has a surface temperature of about –292°F (–180°C). Instead, the liquid would be a hydrocarbon such as ethane or methane, chemicals that exist as gases on the Earth and are extremely flammable (but of course on Titan there is nothing to ignite them). So when scientists designed a probe to land on Titan in 2005, they made sure it would survive a liquid landing.

Also exciting were the methane clouds seen by observers operating the Keck Telescope on Mauna Kea. An example of moving clouds

10 Dec. 2001          11 Dec. 2001          28 Feb. 2002

FIGURE 7.5 Infrared images of Titan taken from Keck Observatory by M. Brown, A. Bouchez, and C. Griffith over an 80-day period showing methane clouds forming and changing in the South Pole of Titan. Also visible are bright and dark areas on the surface, providing a tantalizing backdrop to the *Cassini* spacecraft's exploration of Titan, which began in mid-2004. Courtesy Michael Brown.

that even suggest the development of weather is shown in Figure 7.5. Liquid pools, clouds coming and going on this alien world so far from home: these were our first hints at the Earthlike nature of Titan.

The lead-up to the insertion of *Cassini* into orbit around Saturn on July 1, 2004 was a suspenseful waiting game with almost no data to analyze. As we bore down on the beautiful ringed planet, we had some images of Titan as a mere dot, a single eye peering back at us, and I was lackadaisically analyzing them. I couldn't get too serious, as I was dreaming of the data about to come in. We weren't going to get a targeted flyby of Titan until October 26, but we scientists convinced the flight engineers to let us get a "sneak peek" of Titan on July 3. It was about 200,000 miles away, far more distant than the *Voyager 1* flyby, and not much better resolution than that attainable from Earth. But we had the new "eyes" of ultraviolet and infrared instruments, and we didn't have to look through the Earth's atmosphere. Engineers don't like to take "science data" during critical engineering phases such as orbit insertions or spacecraft trajectory changes: in Chapter 8 we will see how cautious they were during a similar serendipitous flyby of Iapetus. The engineers' conservatism was also a replay of what happened when *Cassini* made its last flyby of Earth, when the

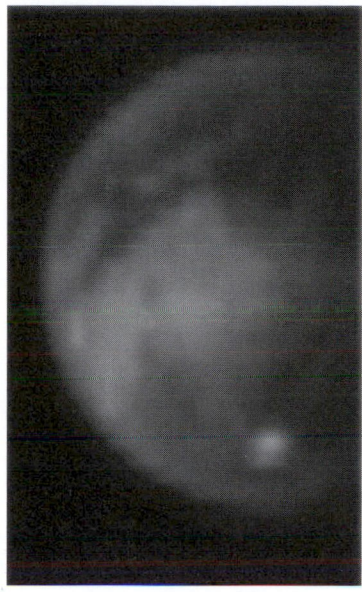

FIGURE 7.6 Our "sneak peek" of Titan in 2004. NASA/JPL-Caltech/University of Arizona. Image processing by Tom Momary. See plate section for color version.

spacecraft made two loops around the inner Solar System to build up the velocity to get out to Saturn: we argued that we should observe either the Earth or the Moon to test and calibrate our instruments. The engineers were adamant in just focusing on the gravitational assist of the Earth flyby as opening up the magnetometer boom would cause the spacecraft to wobble around a bit. We argued and argued and the engineers finally relented and let us observe the Moon. Infrared observations of water ice absorption bands obtained during *Cassini*'s lunar flyby became part of the key proof for water ice at the poles of the Moon.

Again *Cassini*'s fine engineers relented and let us get our glimpse of Titan. We eagerly waited for data to come down through the radio telescopes of NASA's Deep Space Network during July 4, 2004, and scientists working with data from the visual infrared mapping spectrometer (VIMS; I am on the Science Team of that instrument) created the first clear view of the moon from *Cassini* (Figure 7.6). The south polar clouds were visible, as was the thick haze, and tantalizing indications of bright and dark regions on the surface. The bright region had already been detected in Earth-based telescopes: scientists gave it the fanciful name Xanadu, the summer palace of Kublai Khan, popularized by Samuel Coleridge's opium-induced poem Kubla Khan. The tradition has held: regions on Titan are named after worlds of fantasy. But on the surface of this fantastic world, we still couldn't see any firm evidence for lakes, oceans, or ponds. These early views weren't even a hint of what was to come, in the 44 targeted flybys of the nominal mission, some of them swooping in to just 600 miles from the surface of Titan. By the end of the mission in 2017, there will have been over 120 close flybys of Titan.

The first suggestion of a hydrological cycle on Titan, evidence for an interconnected system of streams draining into lakes or oceans, which then evaporated into clouds that rained and fed the streams, came from the *Huygens* probe. *Huygens* was a lander built by the European Space Agency with six instruments including a camera (built by the United States) and chemical analyzers. The probe was attached to the *Cassini* spacecraft and jettisoned from it on December 25, 2004. It landed on Titan on January 14, 2005, two weeks after the serendipitous flyby of Iapetus (Chapter 8). During its descent a camera sent back remarkable images of a drainage system on Titan: streams bleeding into low-lying areas that looked like oceans but weren't (Figure 7.7). An analysis by Ralph Lorenz of Johns Hopkins University's Applied Physics Lab suggested that the probe landed in an area with the consistency of crème brûlée. The probe studied Titan's atmosphere as it descended, and upon landing, another camera sent back images of the surface of Titan with rocks seemingly eroded smooth by some liquid, or perhaps by wind (Figure 7.8). It was a scene not unlike a dry terrestrial stream bed. But we still didn't have proof of a liquid on the surface.

One important point these images do not drive across is how little sunlight there is on the surface of Titan. Because it is almost ten times farther from the Sun than the Earth, Titan receives only about 1% of the sunlight incident on the Earth, and the haze cuts outs out a large fraction of that 1%. It would be a dark world indeed to a person

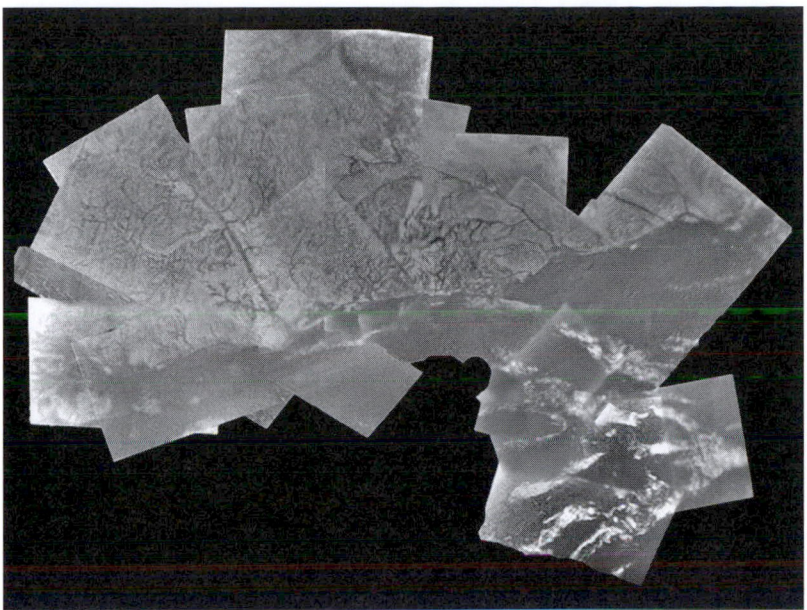

FIGURE 7.7 A mosaic of the surface of Titan from a distance of about 25 miles showing a complex pattern of river channels. The dark regions look like oceans, but they are not. NASA/ESA/University of Arizona.

FIGURE 7.8 The surface of Titan as seen by the *Huygens* lander. Note the rounded rocks. ESA. See plate section for color version.

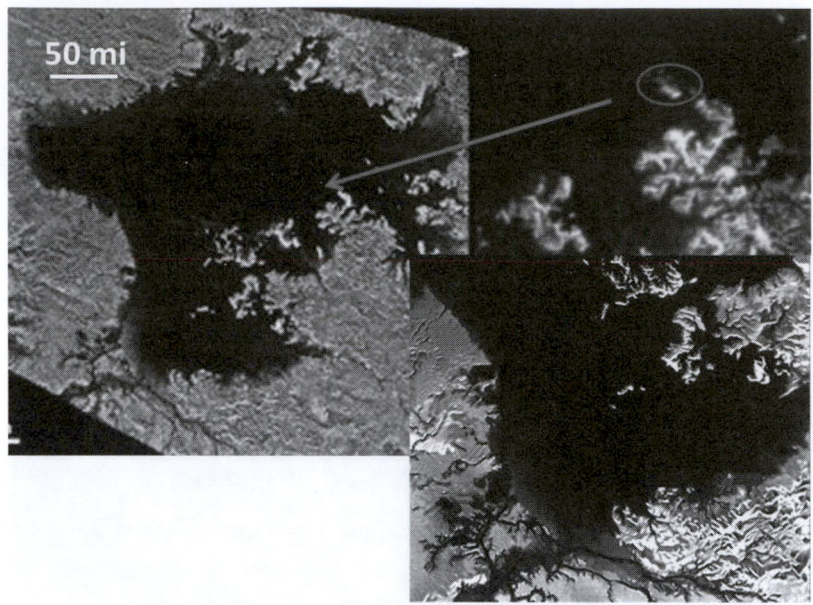

FIGURE 7.9 Radar image of Ligeia Mare, one of Titan's largest seas. Located in the North Polar region, it is about as large as Earth's Lake Superior. The "magic island" (circle) appeared near the top of the large peninsula in 2013. A year later it had almost disappeared. At the lower right is an image of the lower part of Ligeia Mare that has been processed by Antoine Lucas to eliminate noise and bring out the most important features. NASA/JPL-Caltech; processing by Jason Hofgartner.

standing there. Of course, things would be better if you had infrared eyes; if we evolved on Titan, we probably would.

Again radar came to the rescue. Radar was an essential instrument on the *Cassini* spacecraft as it could transmit microwave waves to pierce through the clouds and reflect back clear images of the surface. More than a year after the probe landed on Titan, these images provided the first good evidence that Titan was the only body in the Solar System other than the Earth that had a presently standing liquid on its surface. Figure 7.9 shows a radar image of Ligeia Mare, one of Titan's large north polar seas (mare means sea in Latin). The sea has an extensive system of rivers leading to it, as well as islands, sloping shorelines, and what appears to be the inflow of deposits into the lake. Perhaps most intriguing is the "magic island" (see upper inset) that appeared first in a radar image obtained in July 2013. A year later Jason Hofgartner of Cornell noticed it seemed to have almost completely faded away. What was happening? Were fluid levels changing (unlikely as the shoreline appears to be stable)? Is the island debris that popped up from below or collected on the surface and then disintegrated? Did strong winds or waves erode it away? Signals returned by both the radar instrument and VIMS suggest the seas have only a hint of wave structure (see Figure 7.12 later, for example, which shows a glint from a smooth surface).

One thing of which we are pretty certain is the depth of Ligeia Mare. Some of the radar signal is reflected off the top of the lake, but most of it is transmitted through the liquid – which we think is a mixture of methane and ethane – and bounces back to the spacecraft from the bottom of the sea. To plumb the sea's depth, we measure how much longer it takes for this second signal to be transmitted, and then calculate the distance it had to travel through the lake, i.e., how deep it is. Ligeia Mare (Figure 7.9) is about 500 feet deep, similar to Earth's Lake Superior's average depth of 483 feet. The drainage patterns leading into Titan's lakes look uncannily Earthlike: Figure 7.10 shows a comparison with Lake Powell in Southern Utah. The beauty that characterizes the Canyonlands of Southern Utah may be found on Titan as well as on Mars (see Chapter 3). Images of smaller lakes on Titan are similar to the "Land of Lakes" terrain characteristic of the northern tier of North America from Alaska through most of Canada to Michigan.

Although most of Titan's seas and lakes appear at the North Pole of Titan, the South Pole contains one prominent body of liquid called Ontario Lacus, named after our own Lake Ontario. A vortex of clouds envelops it. The lake seems to be quite shallow, about as deep as a swimming pool (radar data suggests a greater depth of 150 to 300 feet). Like the northern seas, it appears to be remarkably similar to some features on Earth. Figure 7.11 shows a *Cassini* radar image of Ontario Lacus compared with Etosha Pan, a terrestrial salt pan

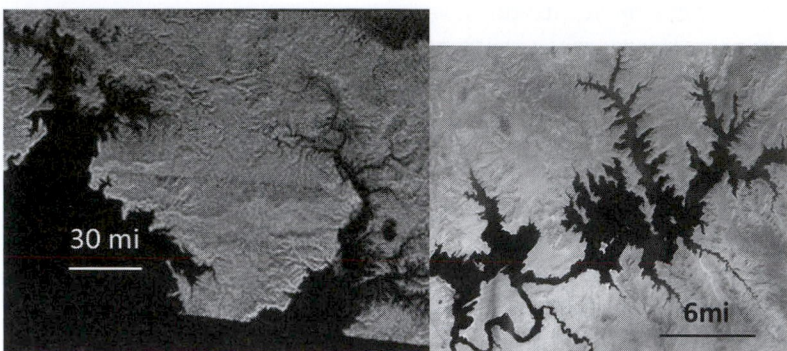

FIGURE 7.10 A comparison of drainage patterns on Titan and Earth. The left side shows a radar image of the northern region of Kraken Mare, the largest sea on Titan, obtained on April 4, 2007. Kraken Mare is three times as large as the largest of the Great Lakes. On the right is a satellite image of Lake Powell in the Canyonlands of Southern Utah. NASA/JPL-Caltech/USGS.

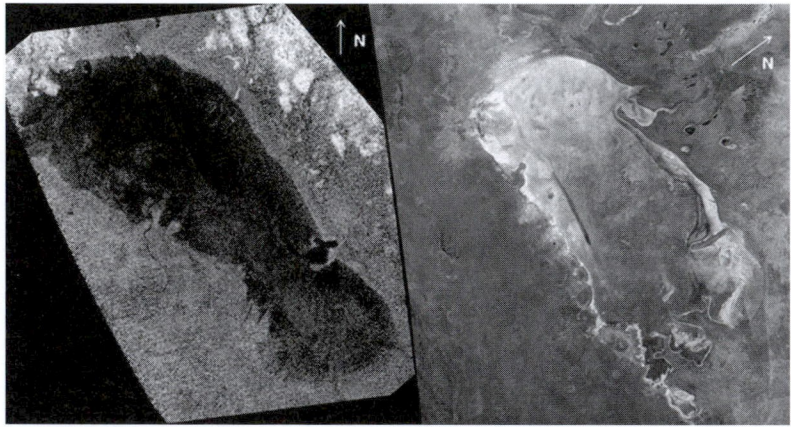

FIGURE 7.11 The left side of this picture shows a radar image of Ontario Lacus near the South Pole of Titan. Drainage channels and associated deltas of transported material (left shore) are clear. On the right is a Landsat image of the Etosha Pan, a terrestrial salt pan, which is a type of dry lake bed. Ontario is 140 by 47 miles while Etosha Pan is 75 by 40 miles. NASA/USGS. Image processing by Thomas Cornet.

in the northern part of Namibia. These features are dry lake beds that fill with a shallow depth of water during the rainy season. But the similarity between Titan's Lake Ontario and Etosha Pan raises a nagging possibility: suppose all these bodies of liquid we are viewing

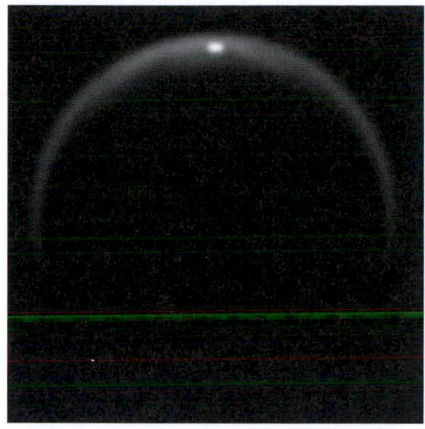

FIGURE 7.12 A glint of sunlight reflected from Jingpo Lacus, near Titan's North Pole, providing strong evidence for a liquid surface. NASA/JPL-Caltech/University of Arizona/DLR.

on Titan aren't really seas and lakes, but are just exceedingly flat areas instead? Frozen ice is an obvious example. Another possibility scientists considered was the smooth dry lake beds, known as playas, that are found in the desert Southwest of the United States and in other arid regions. The radar measurements reflected from both the top and bottom of the seas suggests this analogy is incorrect.

But further positive proof was ours one hot July afternoon in 2009. We were now in the extended mission; the nominal mission ended in 2008. I was sitting at my desk reading email when I opened up an image sent by Katrin Stephan of DLR (Deutsches Zentrum für Luft- und Raumfahrt, the German Aerospace Center in Berlin) to a few members of the *Cassini* VIMS Team. *Cassini* had just executed T58, the 58th targeted flyby, and Stephan was sending around some of the data. I was privileged to be among the first few people to view an extraordinary image, a secret of nature that had just stepped forth for the first time. It was one of those rare moments that make the years of toil in the life of the scientist worthwhile. Our spacecraft had caught a blazing reflection of sunlight from the northern lakes of Titan: it appeared to be the glint of a great polished stone in the sky (Figure 7.12). I abandoned the logical thinking and measured reaction typical of scientists to just gasp in awe at sunlight running along an alien ocean just as it does on Earth. Now we knew the dark features at the North Pole were almost certainly liquid. The sheer volume of liquid hydrocarbons on Titan may be hundreds of times that of the Earth's oil reserves. In case you get any ideas about getting rich from oil wells in space, consider the immense cost of extracting and transporting all that carbon back to Earth.

Over the many years of the *Cassini* mission we have been watching how clouds and lakes evolve on Titan. A full seasonal cycle on Earth is one year; during that time summer and winter come and go in each hemisphere, with the waxing and waning of the terrestrial polar caps. Because its rotational axis is tilted with respect to the Sun, Titan too has seasons, except one full seasonal cycle is nearly 16 Earth years. What kinds of seasonal changes happen on Titan? There are cloud systems on both poles of Titan. Do the clouds feed the lakes with liquid methane and ethane in a cyclical manner tied to the seasons? When the *Cassini* spacecraft arrived at Saturn in July, 2004, it was northern winter, with the North Pole in darkness. A thick cloud cover enveloped the lakes. As summer approached in the north, and the Sun fell on the clouds, they began to dissipate (see Figure 7.13). Meanwhile, the clouds in the south seemed to be increasing as sunlight there declined. There are also clouds in Titan's temperate zones; they were more prominent in the southern hemisphere when we arrived in 2004 during southern summer, and models predicted that the pattern would reverse with more clouds in the north with the approach of northern summer. It hasn't exactly worked that way, and we're still trying to figure out why (or maybe they are yet to appear).

There is at least one firm example of rain on Titan, and the rainfall is being fed by a cloud system. Elizabeth Turtle of Johns Hopkins University's Applied Physics Laboratory and her colleagues had been monitoring Titan with thousands of images when they finally caught a cloudburst in action (Figure 7.14). The rainstorm happened near the equator – near the *Huygens* landing site, in fact. A vast system of clouds swept across the area, leaving a region of darkened land in its wake. The most likely explanation of this darkening is that it was

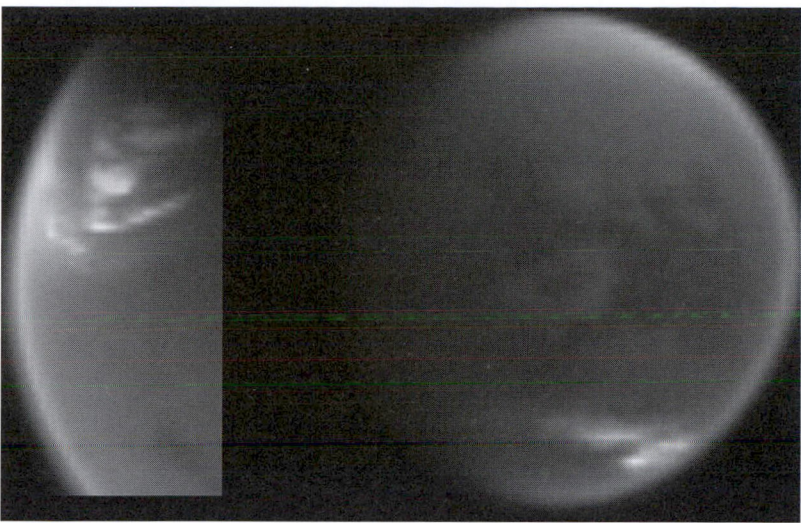

FIGURE 7.13 Clouds on Titan (the colors have been enhanced to bring out cloud features). On the left, the north polar methane clouds are already dissipating as summer approached in May, 2008. The right image obtained in December, 2009 shows an example of clouds in Titan's southern temperate zones. NASA/JPL-Caltech/University of Arizona/University of Nantes/University of Paris. Image processing by Sebastien Rodriguez. See plate section for color version.

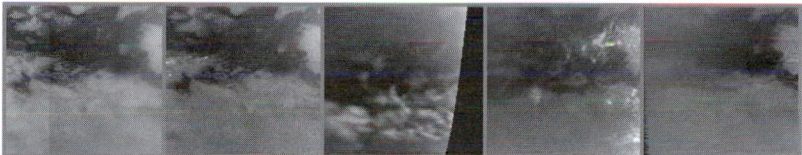

FIGURE 7.14 A cloud burst and rainstorm on Titan. The first picture shows the equatorial region before the action. In the second picture obtained on September 27, 2010, a few scattered clouds appear in the middle left. About two weeks later, the entire bottom half of the image is covered with clouds. By the fourth picture, obtained on October 29, 2010, most of the clouds had dissipated except for a few in the right half of the image, but regions in the center of the image have darkened due to rain. By the last picture obtained in mid-January, 2011, these drenched regions are brightening up again and returning to their original state. The *Huygens* landing site is near the apex of the light-colored triangle that juts out from the left edge of the image. NASA/JPL-Caltech/Johns Hopkins University Applied Physics Laboratory. Analysis done by Elizabeth Turtle.

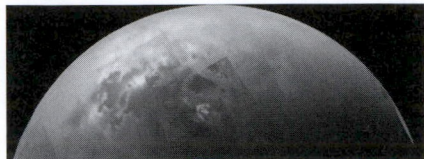

FIGURE 7.15 A *Cassini* visual infrared mapping spectrometer (VIMS) observation of the North Pole of Titan showing that the lakes (the darkest areas) are rimmed by what appears to be a deposit caused by evaporation of the lakes (the bright spot to the northwest of center is another specular reflection; the bright pinkish feature in the center is a cloud). NASA/JPL-Caltech/University of Arizona. See plate section for color version.

caused by drenching rains. The region later brightened to return to its initial state. A team led by Paul Dalba of Boston University found good evidence for additional rain showers.

One question that is on my mind, and that others have often asked me is, is there lightning and thunder on Titan? We haven't found any. Lightning is caused by the discharge of an electrostatic force in the atmosphere, and thunder is the resulting supersonic shock wave. Perhaps we have not yet seen lightning on Titan; it was years before we saw rain. Or maybe we don't know where to look: the chemical reactions that result from lightning may be vastly different in Titan's alien atmosphere; in that case we are not looking at the right wavelengths.

If clouds are feeding the lakes, do the lakes return their liquid to Titan's skies through evaporation? Are the shorelines of the lakes receding? Ontario Lacus seems to be encircled by a collar of evaporated deposits, but *Cassini* scientists cannot agree on whether it is currently changing its area and depth. A VIMS mosaic of enhanced images obtained at the North Pole shows that the lakes there seem to be surrounded by a ring of very different material, sort of a sludge or salt or frost left over after the liquid receded (Figure 7.15). But we don't yet know the composition of the ring. One misconception about the work of scientists is that as soon as we obtain data, it tells us our answers. Often there are years of toil not only reducing the data, but modeling it and comparing it to laboratory samples. In the case of trying to understand the composition of the evaporate deposits around Titan's lakes, we need laboratory comparisons of every material that the deposits could possibly be. We may not have yet measured the right material. The laboratory measurements are exceedingly difficult, as the samples have to be grown in a vacuum and are very tricky to construct (plus some of the substances are poisonous or explosive). My colleague Roger Clark of the Planetary Sciences Institute has been spending decades carefully amassing lab data to compare to planetary surfaces, and he believes he has only measured a small sample of the possibilities.

Earth's evolution has been dominated by the hydrological cycle: the process of water successively existing in three forms of matter, solid (ice and snow), liquid (oceans, lakes, rivers, rain, and water "under the Earth"), and gas (water vapor) to sculpt and erode the Earth, to provide the lubricant that drives geologic processes, such as water erosion and plate tectonics, and to provide the substrate for life itself. Water works because the surface temperature of the Earth is very near the point at which water can exist as a solid, liquid, or gas (the "triple point" in the jargon of the scientist). With a triple point near the temperature of Titan's surface, methane could provide the same function for Titan that water does on Earth. With methane clouds feeding lakes, streams, and rivers that in turn evaporate to a gas that replenishes the clouds to rain down again, Titan appears to have a methanological cycle similar to Earth's hydrological cycle. But under the surface of Titan, scientists have found a liquid water ocean, an ocean that could harbor primitive life forms.

Besides liquid water, life (at least as we know it) needs something to eat, and some form of energy. The orange color of Titan was noted by Gerard Kuiper, but it was Carl Sagan and Bishun Khare of Cornell University who explained it: Titan's surface and haze layer are filled with reddish hydrocarbons, molecules that are the building blocks of life. The nitrogen and methane in Titan's atmosphere are bombarded by solar radiation and polymerize into complex organic molecules that make up the moon's haze layer. These orange-colored haze particles rain down on the surface of Titan to form a thick layer of particles on the substrate of rock-solid water ice. Sagan and Khare created some of these organic molecules in the lab and called them "tholins" (from the Greek word for sepia ink, a common drawing material from ancient times until the nineteenth century) or "intractable polymers," molecules that are the precursors of life. Titan is a celestial laboratory undergoing the same chemical reactions in space that Sagan and Khare performed on Earth. Besides sunlight, Titan has some form of energy: the dissipation of tidal heat from the enormous gravitational pull of Saturn. All three elements for life are there – a liquid water ocean, organic molecules for food, and energy. Some scientists have boldly suggested that liquid hydrocarbons could provide a basis for life in the same way that water does on the Earth, with hydrogen and acetylene (more technically known as ethyne) providing a metabolic pathway for "life as we don't know it." Of course, we haven't found life on Titan, but if we do we will let you know right away.

The constant destruction of methane by sunlight in Titan's atmosphere means the gas must have an ongoing source. The most likely supplier is volcanoes. Even volcanoes on Earth belch out massive amounts of methane. We were pretty excited about seeing some volcanoes actively erupting on Titan, and we have looked very hard for over ten years. There is nothing yet to report, but the radar mapper has found a structure called Sotra Facula that is likely an extinct volcano. Paul Dalba and I searched *Cassini* images for signs of heat in this feature, but we found none: if it is a volcano, it is extinct or dormant. Another region called Tui Regio, in the southwest region of Xanadu, appears to have recent lava flows.

*Cassini* hit the jackpot in finding one common terrestrial feature on Titan: dunes. Dunes are wind-blown features that are found in the dry places of the Earth. We saw they are abundant on Mars, because the Red Planet has dry deserts and wind. During the early flybys of Titan that were dedicated to radar mapping, expansive stretches

FIGURE 7.16 A comparison of dunes on Earth and Titan. The top shows dunes in the Namibia Desert in Africa. The bottom is a *Cassini* radar image of dunes on Titan. NASA/JPL-Caltech.

of strange linear features that the team dubbed "cat scratches" were observed in the equatorial regions of Titan. (Cat lovers were pleased at the early terminology of the *Cassini* mission: tiger stripes on Enceladus and cat scratches on Titan.) Under higher resolution it became apparent that these features were dunes (Figure 7.16). They were found only around the equator, with its vast areas of deserts, and they were collections of wind-blown sand. But unlike the Earth, this sand of Titan's dunes is composed of the organic molecules created high in Titan's atmosphere, providing a large reservoir for this material. Jani Radebaugh of Brigham Young University and her colleagues have mapped wind directions on Titan by noticing the directions in which the dunes point. From this data they have derived the circulation patterns of Titan's atmosphere, which in general are eastward.

Titan is also like the Earth in that it has very few impact craters. Both Earth and Titan are pummeled as frequently as their neighbors, the Moon (in the case of the Earth) and the highly cratered smaller moons of Saturn (in the case of Titan). What makes Earth and Titan different is erosion, especially fluvial erosion caused by liquids on the surface that erase craters. Earth's giant craters are also gobbled up by plate tectonics; we're not yet sure if there are other geologic processes on Titan that erase craters. There are only a handful of verified craters on Titan (about 60 to 70 or so) and another few dozen circular features

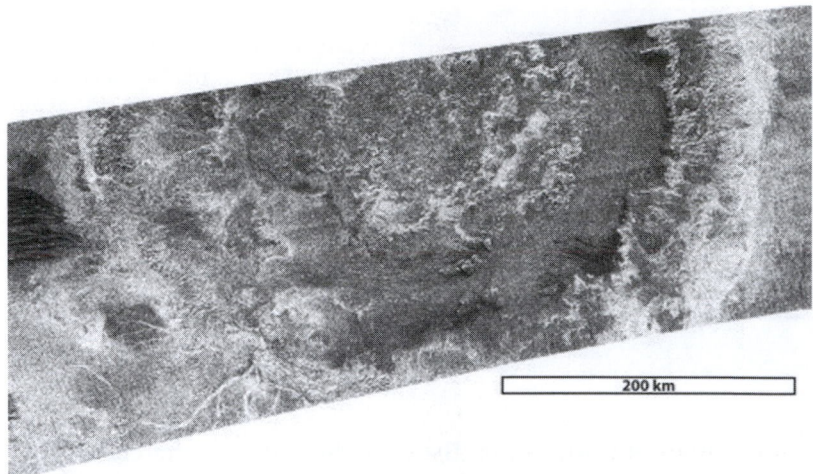

FIGURE 7.17 A radar image of Menrva – a giant impact crater on Titan. Its walls are heavily eroded by fluid flow, and dunes appear in the crater wall (lower right), as well as near the upper left of the image (these dunes appear to lie outside the crater). 200 km is about 120 miles. NASA/JPL-Caltech.

that are suspected of being caused by impacts. The ones that do exist are eroded, with their crater walls washed away by rivulets of liquid hydrocarbons, and dunes marching through the failing walls into the inner sanctum of the crater's floor (Figure 7.17).

The study of Titan is unfolding – just like our exploration of Mars. Astronomers of past generations started by studying the pin-point of light, from which they cleverly deduced a few spare facts. Our generation is like the scientists poring over images from the *Mariners* that first reconnoitered Mars. We will advance to a rover that will bring Titan into its own as a geologic world, and then to balloons and planes in the atmosphere and boats on the seas. The final step – beyond the lifetime of anyone alive now – will be to know the beauty of the place. On that day we will know that Titan is a world that contains works of creation as marvelous as those on Earth, but not of the Earth.

As we leave Titan to explore other worlds, I want to close with a picture that I first viewed as an adult, but that pulled me in

FIGURE 7.18 Frank Paul's "Golden City on Titan" from the back cover of *Amazing Stories*, November, 1941. See plate section for color version.

the same inexorable way that Sy Barlowe's drawing of the Venusian swamp did when I was a child: "Golden City on Titan" by Frank R. Paul, an Austrian–American artist, which first appeared in the pulp science fiction magazine *Amazing Stories* in November, 1941 (Figure 7.18). The begoggled bees that inhabit Titan are lucky to live in a city filled with classic art-deco buildings fueled by abundant hydrocarbon resources puffing up into the atmosphere. And in the background there is a volcano. There may not be any golden cities on Titan, but Paul had a few things right.

# 8 Iapetus and its Friends: The Weirdest "Planets" in the Solar System

I often wonder what it would be like to stand on the surfaces of the celestial bodies that I study. I've imagined skiing on Enceladus, or sitting on the edge of a fissure in the midst of that moon's tiger stripes, with its plumes gushing and sparkling in the sunlight, much as I waited on a bench near Yellowstone's Old Faithful to see its glorious, predictable eruption. One moon's surface I have trouble visualizing is that of Iapetus, a large moon of Saturn. With a diameter of just under 980 miles, it is the third largest moon in Saturn's family, after Titan and Rhea. One side of Iapetus is covered with fairly fresh ice – similar to what you might see on the streets of New York City a day or two after a snowfall. The other side is pitch black – as black as tar or coal. Planetary scientists have long debated how the face of this moon acquired its strange countenance.

Iapetus was discovered by our old friend Giovanni Cassini, in October of 1671. He had just become director of the Observatoire de Paris, at the invitation of King Louis XIV (the "Sun King"), where he stayed until his death in 1712. Cassini noticed that he could see the moon only when it was on the east side of its orbit around Saturn. He stated that Iapetus has "a period of apparent Augmentation and Diminution, by which period it becomes visible in its greatest Occidental digression, and invisible in its greatest Oriental digression. It begins to appear two or three days before its conjunction in the inferior part and to disappear two or three days after its conjunction in its superior part."[1] Cassini correctly deduced that Iapetus keeps the same face toward Saturn, in the same way that our own Moon is locked toward the Earth, and that one side would have to be very bright and the other very dark for Iapetus to be visible on only one side of its orbit (see Figure 8.1). Cassini did not observe the dark side

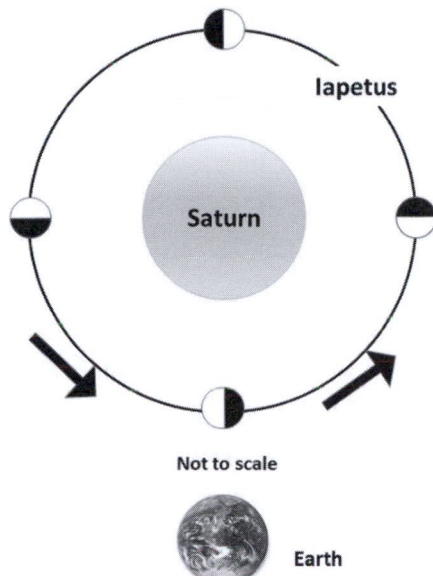

FIGURE 8.1 Like our own Moon, which keeps the same face toward the Earth, Iapetus keeps the same face toward Saturn. One half of the moon, the "leading side," which is the front of the moon with respect to its motion around Saturn, is very dark, while the "trailing side" is bright. An observer on Earth would see a bright Iapetus when it is on the east (right) side of its orbit, and a dim Iapetus when it is on the west (left) side. Iapetus is more than six times brighter on the east side.

of Iapetus until 34 years later, with one of the ever-larger refracting telescopes he kept building at the Observatoire. Through a telescope, Iapetus appears over six times brighter when it is on the east side of Saturn.

But Cassini went on to wildly speculate, and he made the following comparison to Earth: "the globe of the satellite has some diversity of parts analogous to that of the Earth, the one part of whose surface is covered by the sea, which is not so fit to reflect from all parts the light of the Sun, as the continent which makes up the other part."

Cassini named the four saturnian moons he discovered – Tethys, Dione, Rhea, and Iapetus – the Sidera Lodoicea after King Louis XIV, his patron and supporter, much the same as Galileo had originally named the four satellites he discovered after the Medici. Its present name was suggested by John Herschel in keeping with mythological themes for each planetary body: Saturn's moons were named after the Greek Titans, the brothers and sisters of Chronus (Saturn), which include Hyperion and Phoebe, as well as Cassini's four original moons,

Enceladus, Tethys, Dione, and Rhea. Some scholars have connected Iapetus with the biblical Japheth, one of the three sons of Noah – the others are Ham and Shem – who were the progenitors of all the peoples of the Earth. The ancient Greeks gave lineages to their gods, but the Hebrews gave a lineage to everyone.

The first sensible idea explaining the origin of the dark side of Iapetus was given by Steve Soter, a research scientist at Cornell University. Speaking at a conference on satellites at Cornell in the August of 1974, Soter hypothesized that meteorites impacted the surface of Phoebe, the outermost known moon of Saturn at that time, and kicked up particles that escaped the moon. The orbits of these particles decayed to cause them to spiral in and strike the hemisphere of Iapetus that led its motion around Saturn (the "leading side," in astronomical terms), much the same way a car picks up bugs and dirt on its front windshield and grille. Soter's idea clearly explained why it was the leading side of Iapetus that was so dark.

But some geologists did not like an astronomical explanation for the strange face of Iapetus. They thought a more likely cause was a giant volcanic eruption of dark material onto its surface. Again, scientists favor explanations that call on phenomena in their own field. When an astronomical explanation involving the impact of an asteroid or comet on the Earth was offered for the extinction of the dinosaurs – and most other terrestrial species – about 65 million years ago (Chapter 4), geologists held onto their ideas of climate change prompted by massive volcanic eruptions in the Deccan Indian flats. But it is this advocacy of disparate hypotheses that drives the collection of more data within the marketplace of scientific ideas.

The two leading theories of the origin of the dark side sparred for decades, driving observations and prodding scientists to design missions to go there. Observing just a tiny orb in their instruments, astronomers deduced that Iapetus had water ice on its surface, and that its dark side was rich in hydrocarbons, the building blocks of life. Water appears very bright to our eyes, which are sensitive to visible wavelengths, but in the infrared, water is very dark at certain

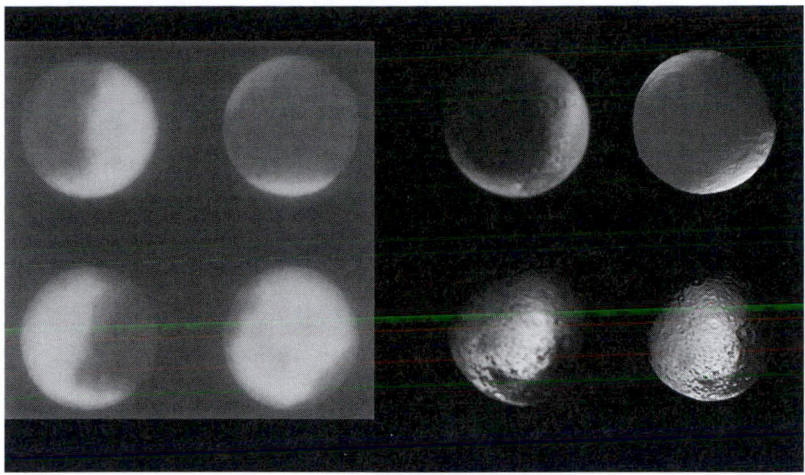

FIGURE 8.2 The left four globes show Morrison et al.'s map of Iapetus (viewed at four different aspects), based entirely on telescopic data prior to any spacecraft visits. The right four images show comparable data obtained by the *Voyager* (bottom) and *Cassini* (top) spacecraft. David Morrison and NASA/JPL-Caltech.

wavelengths, especially near three microns (corresponding to wavelengths almost ten times as large as visible light). Astronomers made measurements of the bright side of Iapetus at many wavelengths, and it was dark in all the places water ice is dark. Hydrocarbons, the building blocks of life, were hypothesized to exist in the dark material because it was reddish. The simple chemical groupings that comprise hydrocarbons become polymerized, dark, and reddish through time, just like the dark side of Iapetus.

In another observational feat, David Morrison, now at NASA Ames Research Center, led a team that put together observations of Iapetus at various locations in its orbit, and in 1975 they published a map of its reflectivity (Figure 8.2). Astronomers have long deduced all sorts of things about planets, moons, stars, and galaxies from tiny pinpoints of light and colors appearing in their telescopes and on their computer screens. We have derived the composition of the surfaces of planets and moons, mapped bright and dark patterns, and figured out how rough or porous surfaces are. I often think that is what science is: making discoveries from nothing, or at least something that looks like very little or even a big mess.

What magic did Morrison and his team perform to make this map from a pinpoint of light? He started with gathering measurements of the light curve of Iapetus, which expresses how the brightness of Iapetus changes as we look at different places in its orbit, corresponding to different geographical longitudes. Morrison figured out what brightness pattern Iapetus would have to have on its surface to have its observed light curve. Morrison's modest conclusion to his team's superb piece of scientific sleuthing was that "We offer no explanation for the unique photometric properties of this satellite."[2] Morrison did not cite Soter's work – even though he was at the conference in 1974 – probably because Soter never published his work in a peer-reviewed journal. Peer review is the "gold standard" of scientific discovery. Scientists write up their results as a paper that they submit to a scientific journal, either one of general interest such as *Science* or *Nature*, or one in their own field, such as *Icarus* or the *Journal of Geophysical Research* for planetary sciences. Other scientists read these papers for obvious mistakes, weaknesses in experimental protocol, unsubstantiated claims, or lack of originality. Soter's idea did not at first go through the process of peer review, although he is widely credited with being the first scientist who proposed the model. Although Morrison's map is not unique – he pointed out that different patterns could yield the same light curve – it is uncannily similar to the first spacecraft images of Iapetus returned by *Voyager 1* at a distance of about a million miles. Later images returned by *Cassini* show an even greater resemblance (Figure 8.2).

My own early work on Iapetus was done in collaboration with a very famous person – Carl Sagan – and a person who would become famous later – Steve Squyres – as the Principal Investigator of the Mars Exploration Rovers *Spirit* and *Opportunity*, as well as my mentor and thesis advisor, Joe Veverka. Sitting together at Cornell's Spacecraft Planetary Imaging Facility (SPIF), we precisely measured the reflectivity of the surface of Iapetus from *Voyager* images. We showed that in its very darkest regions, it reflected only about 2% of sunlight that fell on it, making it as black as the soot from a candle. The brightest areas of the moon reflected about 70% of sunlight, similar to that reflected by slightly dirty snow. These numbers were much more extreme than previously thought. Some work I did later with Joel Mosher suggested that the change from dark to bright was very gradual, evidence that we felt strongly suggested a "Soter-like" origin to the dark side of Iapetus. A volcanic origin would cause a very distinct, abrupt change in brightness between the bright and dark regions, as the extrusion of a dark lava-like material from the interior of Iapetus would flow over the surface and come to a distinct stopping point. Later evidence would show that we were not entirely correct about the gradual change.

It turns out we *Cassini* scientists got a close look at Phoebe even before we flew by Iapetus. Phoebe is one of the outer moons of Saturn, the largest one. It was discovered by the great American astronomer William H. Pickering (1858–1938) in 1899 as the ninth moon of Saturn and the first moon to be discovered on photographic plates. One hundred years later astronomers using sensitive CCD cameras – not unlike those in your cell phone – and large telescopes began to discover the other outer moons of Saturn, which group into three classes based on their orbital characteristics, and are named after mythological figures in Norse, Celtic, and Inuit mythology. There are a total of 37 known so far, with names like Paaliaq, Skathi, Bebhionn, Jarnsaxa, Suttengr, and Bestla, not as familiar as the names of Titans that came down through Greek mythology. Phoebe was the very first moon we saw close up in Saturn's system: years earlier mission planners on the *Cassini* Project made sure that Phoebe was in the right place in its orbit as *Cassini* crossed it. Phoebe is so far from Saturn – about a tenth of the distance between the Earth and the Sun – that it takes over a year and a half to orbit Saturn. Assuring that *Cassini* and Phoebe were in the same place at the same time (or nearly in the same place – *Cassini*'s closest approach on June 11, 2014 was 1,285 miles) was tricky.

As *Cassini*'s seven-year cruise to Saturn was nearing its end, hardly spoken was the possibility that we wouldn't ever get into orbit around Saturn, that all we would have would be some infrared images of a distant Titan and a nice encounter of Phoebe. This encounter occurred before Saturn Orbit Insertion (SOI), the precise engineering feat that captured *Cassini* into its embrace around Saturn. As we approached Saturn, I was quietly analyzing infrared images of Titan – which was just a point of light – much as a ground-based observer might do but in infrared wavelengths that had never been seen before; I even published a peer-reviewed paper on this work. But then we started closing in on Phoebe. The resolution didn't get better than *Voyager*'s until about 12 hours before closest approach: by that time *Cassini* had already gone into its automatic mode of executing observations. We had to wait until it had finished its work at closest approach, turned toward the Earth, and downlinked its data. Just this last event takes eight hours.

So on June 12, 2004, images and other data on Phoebe began to be returned by *Cassini*, more than three weeks before SOI. We saw a battered moon, saturated in craters that would have formed the dust that escaped the moon throughout the ages (Figure 8.3). Some of these craters may have been formed by members of the family of Saturn's outer irregular moons: many such small undiscovered bodies that sometimes collide with each other should still be out there. The largest craters on Phoebe have bright icy rims and layered deposits. I've thought that standing in one of those craters would be like looking up from the bottom of the Grand Canyon. Some of the craters are conical, which is more characteristic of underground blasts rather than impacts. What could have caused them? Pockets of carbon dioxide or another gas exploding as it sublimates? Some *Cassini* scientists believe Phoebe formed in the Kuiper Belt (see Chapter 9) and was captured by Saturn as it drifted inward. If this idea is true, it would have formed in a colder place and have been able to sustain large pockets of ices other than water ice. Indeed, carbon dioxide ice was discovered during the flyby.

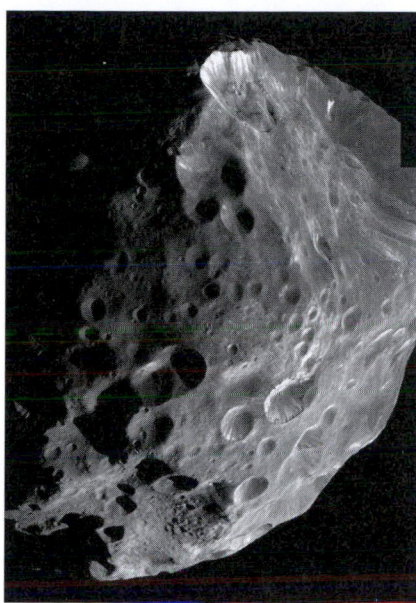

FIGURE 8.3 A *Cassini* image of Phoebe obtained on June 12, 2014. The moon is 136 × 127 miles. NASA/JPL-Caltech.

So after this first encounter by *Cassini*, I could imagine standing on the surface of Phoebe, but I still couldn't imagine what the surface of Iapetus looked like if I were standing there. The brightest material is almost 40 times as bright as the darkest material (even though through a telescope it's only six times brighter on one side). What on Earth exhibits such changes in dark and bright? A tar pit in the middle of a snow field? I had been to Hawaii, so I could envision the edge of a volcanic lava flow. It just stopped. And of course it looked different from the underlying land it had engulfed. But I never accepted the geologic explanation for the dark side, and I thought it changed gradually from bright to dark. But what did a gradually changing surface look like? Was it just a transition to ever cleaner snow, similar to what one might see in a drive from the dirty slushy streets of New York out to the pristine snows of the surrounding countryside? Or were there small dark patches on Iapetus that got bigger and bigger toward the dark side?

The *Cassini* mission would bring us to the next phase of studying Iapetus, and observations from the spacecraft would help us visualize the surface of this moon. Instead of viewing it from *Voyager*'s distance of one million miles away, we would get to see Iapetus from a distance of only one thousand miles. This is the difference between viewing your pet cat on your lap and viewing him from a thousand feet away, where he would not even be recognizable as a cat. The only problem is, we had to wait until 2007, three years after *Cassini* would start its orbital tour of Saturn. Iapetus is far away from *Cassini* – more than nine times as far as the Moon is from Earth – and it takes a lot of spacecraft fuel to get out there. And we wanted to get some close views of the inner moons before making that journey. Luckily, there was a serendipitous, unplanned flyby of Iapetus on New Year's Day 2005, during the period *Cassini* was making its set-up trajectory to deploy *Huygens*, the European probe that was to land on the surface of Titan.

We *Cassini* scientists were thrilled at this opportunity of a "sneak peek," but there was a problem. The European *Huygens* probe was to be deployed from *Cassini* on December 25, 2004, and land on the surface of Titan on January 14, 2005. The conservative, careful engineers on *Cassini* did not want to take any scientific data during this period, as they didn't want anything to screw up the landing on Titan. Amanda Hendrix of JPL and I led a group of scientists who proposed to observe Iapetus during this untargeted flyby. We argued the scientific payoff was worth the small risk. We would get a much earlier view of this enigmatic object, enabling us to pose new and better questions for the real flyby, and we would see regions of the moon not visible during the 2007 encounter. The position of the Sun was also different, allowing us to see the shadows of craters and mountains pop up on the surface. I remember arguing before a *Cassini* Science Project Group Meeting in Lisbon in June 2002 what a great scientific payoff this flyby would give us, and Steve Squyres, a member of the imaging team, sitting in the back of the room, was smiling faintly, thinking perhaps back to those days in graduate school when we worked together on this weird moon (Carl Sagan had passed away by then, but Joe Veverka was still going strong).

Bob Mitchell, the pro-science, very reasonable project manager, who has that rare skill among engineers to know just how much risk to safely take, gave us the go ahead. We would get to within 36,000 miles of Iapetus, about 15 times closer than *Voyager*. We won, but the most careful, risk-averse engineers got the last word. During a review of the *Huygens* landing phase of the mission, Gentry Lee, a JPL engineer with a superb ability to foresee any and all problems, noticed a potential issue that had been overlooked. The mass of Iapetus wasn't very precisely known, and if it were more massive than expected, its larger gravitational pull might pull the *Huygens* probe off the orbit intended to land it on Titan. We thought this scenario was highly unlikely, but the project didn't want to take any risks during this critical part of the mission, especially since we were delivering a European lander. We simply couldn't fail our European colleagues. A compromise was reached: we would fly the spacecraft to within 72,000 miles of Iapetus instead of 36,000 miles. This distance would still give us seven to eight times the resolution of *Voyager*, 15 times if the better resolution of the *Cassini* camera is taken into account. And, of course, the spacecraft could see with all those infrared and ultraviolet "eyes" that *Voyager 2* didn't have.

Over the 2005 New Year's weekend, the data trickled into JPL. We arranged to view the images as soon as they were constructed by JPL's image processing lab. My favorite image from that anxious wait on the second floor of JPL's Space Flight Operations Facility is one of the "walnut" ridge that extends most of the way around the equator of the moon, and reaches 12 miles high, taller than any mountain in the Solar System (see Figure 8.4). At times it breaks into a chain of discrete peaks. The ridge has been there for a long time, because it is pockmarked with craters. Scientists are not sure how this equatorial ridge formed, but the most likely explanation is that it somehow "froze" into the moon when it was spinning much more rapidly, shortly after its accretion and prior to its being locked in synchronous rotation. There are other fanciful theories, such as the ridge is the collected debris from an ancient ring around Iapetus, or

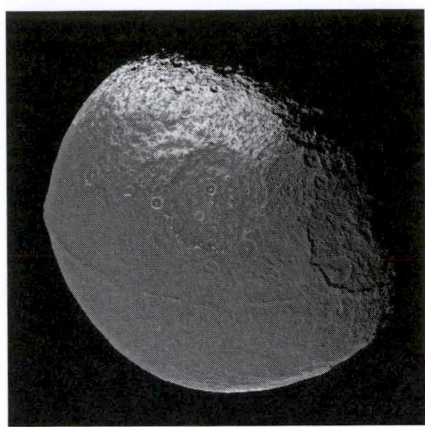

FIGURE 8.4 An image of Iapetus taken by *Cassini* on New Year's Day 2005. The image shows huge impact craters in the dark region of Iapetus, known as Cassini Regio, and a unique and poorly understood "walnut ridge" spanning the equator. NASA/JPL-Caltech.

that it is an upwelling from the moon's interior. Is the ridge connected to the bright–dark dichotomy of Iapetus? We just don't know, and no scientist has yet made a solid connection.

But it was not until the close flyby on September 10, 2007 that I felt I was standing on the surface of Iapetus. This flyby had a closest approach of about a thousand miles. We arranged to have a big party in JPL's von Karman auditorium, and the Imaging Team's Torrence Johnson and I were supposed to comment on the images as they came down. We were upstaged by an "appearance" by the famous science fiction author Arthur C. Clarke. Iapetus was at the core of his novel *2001: A Space Odyssey*. The supremely intelligent aliens who had buried the monolith on the Moon placed another larger monolith on Iapetus. This structure was a stargate, a passageway to the transformation of the astronaut Dr. David Bowman into an immortal Star Child destined to bring Earth to its next stage of development. (In the movie, the climax takes place at Jupiter.) *Cassini* Mission Planning Team leader and space artist David Seal had invited Clarke to our shindig. At 89 years old, he was too frail to travel to California from his home in Sri Lanka, where he had lived since 1956, so he sent us a video to play during the event. The video acknowledged the important role Iapetus played in *2001*, and it described Clarke's pleasure at *Cassini*'s visit to the strange moon. But the real pleasure

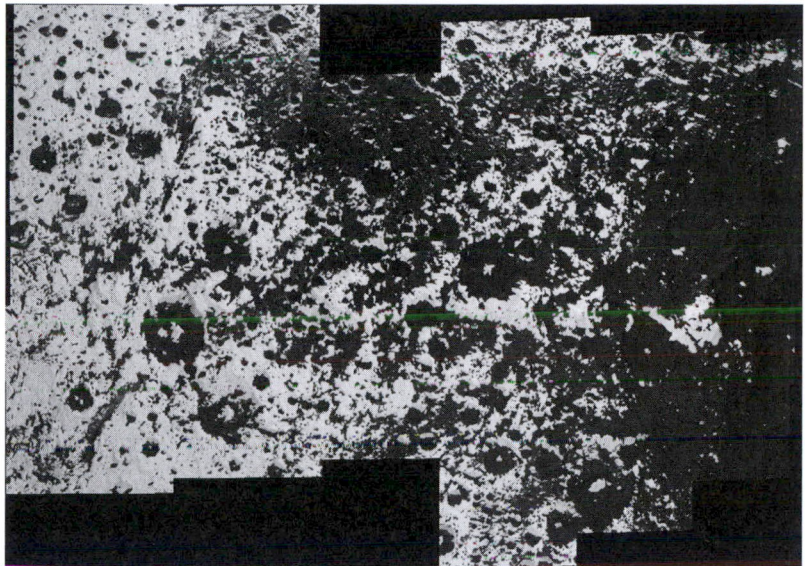

FIGURE 8.5 A close-up mosaic from the Iapetus flyby in 2007, showing the interface between the bright and dark material. From a distance the transition appears gradual, but this closer image shows the discrete patches of dark material becoming more and more rare as one moves from the dark to the bright material. NASA/JPL-Caltech.

was ours, as we listened to this science fiction author we all admired praise our technical prowess for reaching this distant world. (I originally thought Gentry Lee had invited Clarke, as they coauthored the science fiction work *Cradle* and the *Rama* series, but Dave Seal told me he simply emailed Clarke with the address listed on his website, and Clarke responded almost immediately with the videotape. Clarke died in March, 2008.)

After this bit of praise, the images came pouring down from heaven (through the Deep Space Network of NASA's radio telescopes). They were spectacular– and Torrence and I had to stand up in front of hundreds of smart people and give our instant interpretations of them! What really surprised me was the patchiness of the features (see Figures 8.5 and 8.6): they look almost painted onto the surface. John Spencer of Southwest Research Institute and Tilmann Denk of

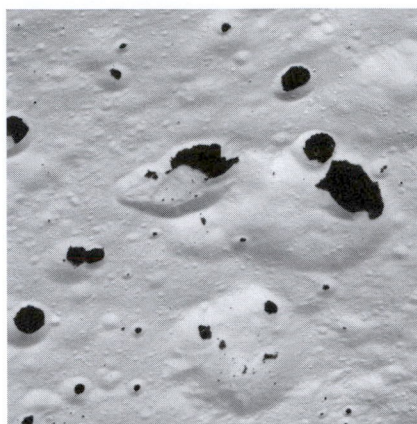

FIGURE 8.6 Tiny patches of dark material toward the bright side of Iapetus. This material persists in the sunlit rims of the craters, where the warmer temperatures have driven ice away. NASA/JPL-Caltech.

the Free University of Berlin later published a model explaining the patches: they were due to "thermal migration." This phenomenon is a feedback process in which darker, hotter patches drive away water ice molecules, which in turn migrate to and get trapped in the colder, brighter areas. The bright areas keep getting brighter as they are colder and can more effectively condense frost, while the dark areas get darker because they are warmer and drive frost away. I was just beginning to see what would be at my feet if I were standing on Iapetus: a bright frosty surface, almost a field of snow, covered with dark patches that grew more numerous toward the dark side.

If there is a single telltale sign to the mystery of Iapetus, and one that clinched the origin of the dark material we were seeing, it would have to be the Phoebe ring, a vast cloud of dust that forms a giant torus around Saturn. This torus orbits Saturn at a distance of a tenth of an astronomical unit: the same position of Phoebe itself. The dust in the rings consists of tiny particles of sooty material that is blasted off the surface of Phoebe by the slow but incessant bombardment of meteoroids. This ring supplies the dark material that strikes the leading side of Iapetus. The Phoebe ring was hypothesized to exist, but wasn't discovered until 2009. It was not detected by *Cassini*'s instruments, but by the *Spitzer Space Telescope*, NASA's infrared telescope in an Earth-trailing orbit about the Sun that is one of NASA's

Great Observatories (the *Hubble Space Telescope* is another one). Even with a Cadillac-like spacecraft such as *Cassini* studying the saturnian system, important discoveries continue to be made from ground-based and other space-based instruments. We've already seen how a staggering 37 small outer moons of Saturn were discovered by Earth-based telescopes. The terrestrial instruments can look at vast sweeps of sky all at once in their search for moons, while the *Cassini* instruments can only see little "postage-stamp" areas as the spacecraft is inside the saturnian system (plus spacecraft time is too valuable to spend on moon searches). It is also true that the large telescopes on the Earth, from the 200-inch "Big Eye" Hale Telescope on Palomar Mountain, to the giant Keck with a 33-foot mirror, are more sensitive than the relatively small *Cassini* telescope.

I spoke recently with the leader of the team that first detected Saturn's Phoebe ring, Anne Verbiscer of the University of Virginia. This key discovery began with a conversation between Verbiscer and her husband, Mike Skrutskie, also at the University of Virginia. They were sitting on their front porch in Earlysville, Virginia, discussing some unusually bright data Verbiscer had obtained on Phoebe. Was it outgassing, or had it been hit by an asteroid or by an unseen small moon? They decided that the brightening was instead just a fluke in the data, but the idea of dust collecting near Phoebe's location – Steve Soter's dust that painted one hemisphere of Iapetus – got Verbiscer and Skrutskie thinking about the idea of using the sensitive *Spitzer* to search for a giant outer ring at Phoebe's orbit. *Spitzer* was designed to look at the hot dusty shells around young stars, the shells from which planets form. Because Phoebe is quite dark – it absorbs about 90% of solar radiation falling on its surface – Verbiscer reasoned that the particles blasted from its surface would also be dark – and hot, in the same way that an asphalt path is hotter than a cement one. Dust particles in the Phoebe ring should be strongly radiating in the thermal infrared and visible to *Spitzer*.

Just as all other scientists are required to do when they want to use one of NASA's Great Observatories, Verbiscer, Skrutskie, and their colleague Douglas Hamilton of the University of Maryland wrote and submitted a proposal to seek out Phoebe's dust ring with *Spitzer*. The proposals are evaluated by a panel of other scientists. During this peer-review process, this group of skeptical experts finds all the weak areas in a proposal and rejects most requests; there is just not enough time to go around. Verbiscer's proposal was accepted in 2006, but the observing periods were to be slipped in only when bits of idle time on the telescope's schedule opened up, and when Iapetus was far from the Sun and easily observable. No observations were made in 2007 or 2008, even with the compelling event of *Cassini*'s Iapetus flyby unfolding in 2007. Finally, in February, 2009, the team was able to get data, only three months before *Spitzer* became unusable for most experiments, as the liquid helium that cooled its instruments ran out. The team discovered the glowing ring, but only after a painstaking stretch of data analysis. The ring itself is tenuous, presenting a gossamer celestial torus with a diameter of about 300 times that of Saturn. The team had to mosaic many images together to reveal the vertical extent of the ring.

There was a *Cassini* Project Science Group meeting in London in June, 2009, and I was organizing a workshop to discuss recent results on icy moons. After I had put together the agenda, Verbiscer asked to have a short period of time to discuss a potential discovery. I penciled in her talk at the end of the workshop, on a Friday afternoon. Verbiscer is a modest unassuming person, so the discovery of the Phoebe ring was described with little fanfare. Sitting in a drab lecture hall at University College London, we all quickly grasped the importance of the results as the carefully analyzed data were shown. At the end of the workshop, we rushed up to congratulate her and to pledge not to publicize the discovery. It had not been peer reviewed, and some prestigious journals such as *Science* and *Nature* do not like to accept discoveries that have already been announced, even if they have not been published. Another danger for Verbiscer's team was that they knew of some earlier images of Saturn's moons from *Spitzer* that showed the ring clearly and that were already archived in a public database. Verbiscer and her team could conceivably be scooped. By August the team had submitted their work to *Nature*, which reviewed and accepted the paper the next month.

Besides the *Cassini* camera, another powerful tool on the spacecraft was its radar antenna. Radar can pierce through the upper layer of Iapetus and tell us the depth of the dark deposits. My colleague Steve Ostro of JPL was leading the effort to gather radar data from the *Cassini* spacecraft. He and his team found that the dark material was only a foot or so in depth, which is not surprising given the low density of the Phoebe ring. Ostro died in 2008, two years before his final scientific paper on Iapetus was published, tragically cut down by cancer way before his time. His colleagues finished writing the paper for him.

But perhaps some of the most significant findings about Iapetus came from VIMS, the Visual Infrared Mapping Spectrometer ("my" instrument: I'm on the science team led by Robert H. Brown of the University of Arizona). There was fun and there was the fundamental. Soon after the sneak preview in January, 2005, I realized if we constructed a map in the infrared at a wavelength of three microns, the bright material would be dark, and the dark material bright. Water ice is dark at three microns, but hydrocarbons are bright, as least brighter than water ice. I asked my JPL colleagues Joel Mosher and Tom Momary to make a map at three microns and compare it to a map in the visible. Figure 8.7 shows the result, a stark representation of the reversal of the bright and dark regions as one moves from the visible to the infrared region of the spectrum.

The more fundamental discovery was made by VIMS science team member Dale Cruikshank of NASA Ames Research Center. One of the original goals of the *Cassini* mission was to search for the elusive dark material in the Solar System, the organic-rich goo that scientists hypothesized provided the building blocks of life. This nutrient-providing, complex material, some of which forms polycyclic aromatic hydrocarbons (PAHs) could not be formed in the inner regions of the Solar System, as it was too hot for its chemical bonds to

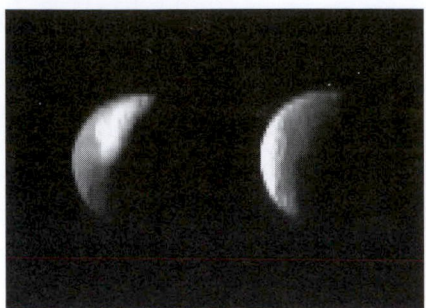

FIGURE 8.7 Two infrared images of Iapetus, one at one micron (left) and the other at 3.5 microns (right). On the left, the polar cap appears bright and the dark side is dark, but on the right the polar caps become dark and the dark material bright. In both wavelengths the light is reflected light and not thermal radiation. NASA/JPL-Caltech. Image processing by Thomas Momary. See plate section for color version.

be preserved. Instead, it was formed in the outer reaches and brought into the inner Solar System by comets (see Chapter 4). Polycyclic aromatic hydrocarbons had been observed in interstellar space, but astronomers just couldn't pinpoint their presence in our neighborhood. PAHs are best seen in the infrared, which isn't easily accessible from Earth but is easily seen by the eyes of VIMS. After nearly ten years of painstakingly searching the VIMS data for the spectral signature of PAHs near three microns, Cruikshank and his colleagues published a paper announcing their discovery. The weirdest "planet" in the Solar System contains the building blocks of life: we may all be connected to Iapetus in the most profound way.

There is another odd moon that seems to be linked to Phoebe and Iapetus: Hyperion, discovered in 1848 and named after the Greek Titan of close observation. Located inside the orbit of Iapetus (but outside that of Titan), Hyperion is the only moon of the giant planets that is known to rotate chaotically. (As we shall see in Chapter 9, the four smallest known moons of Pluto are also in chaotic rotation.) The other moons either keep the same face toward their primaries (like our Moon and Iapetus) or they spin on their axis with a regular period (like Phoebe, which rotates about every nine hours). Hyperion just can't decide what its rotation period is. Mysteriously, it seems to rotate in a well-behaved fashion for a while, but then it goes wild. As a graduate student, I remember presenting *Voyager* results that seemed to show Hyperion had a regular spin period. But soon after my talk, ground-based observers who had been collecting data for a much longer time period than that covered by *Voyager* showed that the *Voyager* rotation rate was wrong. Furthermore, they couldn't determine what it was. A theoretical model developed by Jack Wisdom of MIT came to the rescue and explained its chaotic state. In a simplified nutshell, Hyperion's irregular shape and non-circular orbit allow gravity to consistently "bump" it out of its rotational state. Again, the value of ground-based observations was shown: the long temporal base of the telescopic data trumped the "blink of an eye" presented by a spacecraft flyby. Of course I was very excited to be presenting a paper in front of all these famous scientists – it was one of my first talks – but it was good to be proven wrong so early in my career. After that, I wasn't afraid of being wrong, and I realized that self-correction is essential to the progress of science, as your ideas and results prod other scientists to take more data and develop more theories. This is how science advances.

Voyager images of Hyperion showed its very irregular shape, but a collision with another object may have set off its eternal tumble. Perhaps it was hit millions of years ago by an errant moon or a meteoroid wandering into Saturn's system. Hyperion seems to have the same color as the dark material on Iapetus, but it is not as dark. I wrote a paper in 2002 that hypothesized that the material from Phoebe that Iapetus didn't accrete ended up on Hyperion. Other scientists thought both Hyperion and Iapetus were coated by material from Titan, or that the collision on Hyperion provided debris that was later swept up by Iapetus. It is hard to figure out how material can move outward from Titan, but I believe the discovery of the Phoebe ring makes this point moot.

The veil over Hyperion was raised further after *Cassini*'s targeted flyby on September 25, 2005. Every first close encounter with

FIGURE 8.8 The *Cassini* image of Hyperion obtained during its targeted flyby. The longest dimension of Hyperion is about 216 miles. NASA/ JPL-Caltech.

a moon opens a new door, but this door opened into a world at least as odd as Iapetus. Hyperion appeared to be a giant sponge (Figure 8.8)! By measuring the gravitational pull on the spacecraft, *Cassini*'s radio receiver was able to determine that the density of Hyperion is only about half that of ice, which is about the least dense material in the outer Solar System (rock is typically about five times as dense as ice, and metal is about eight times as dense). The inescapable conclusion

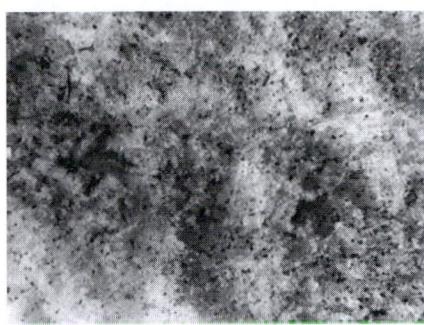

FIGURE 8.9 Cinder dust embedded in a snow bank in eastern Pennsylvania. The hotter cinders form pits in the snow.

is that Hyperion is full of spaces in its interior – that it is an unconsolidated rubble pile held together by gravity. This picture suggests not only that it was once a world in collision, but that it may have broken up completely and then came back together as a collection of smaller bodies. Or it may simply be one remaining fragment of a much larger body. Its odd surface has the outline of a giant crater almost as big as the moon itself. Enormous landslides sprawl down the rims of the crater. The floor of the crater is pockmarked with dozens of smaller deep craters that have very dark bases. How did these craters evolve into these unique structures? One idea is that dark material – perhaps Phoebe dust that wasn't picked up by Iapetus, or material from the meteoroid that smashed into Hyperion – collected in the crater and that this darker, warmer material "melted" down into the ice. Again, thermal segregation would have kicked in: the darker warmer areas would sublimate any ice and drive it off to the colder, brighter areas. On Earth, a similar phenomenon is seen in "suncups," depressions in snow fields often encountered by mountain hikers. Figure 8.9 shows another terrestrial analogy: cinders mixed into a snow bank in Nazereth, Pennsylvania warmed up and drilled holes into the ice over time. This picture was taken about three days after a snow storm.

Before we move onto what might be another Earthlike body in the Solar System – Pluto – it is worth noting that there are two "ecologies" at work in Saturn's system, and both are controlled by large tenuous rings. Feeding the E-ring from its plume and dusting the main inner moons with bright frost, Enceladus controls the ecology of the inner regions of the saturnian system, while Phoebe and its ring of dark dust controls the outer regions populated by Iapetus and Hyperion. There is nothing else quite like it in on Earth or in heaven. As we saw in the last chapter, Titan is Earthlike in its appearance and history, but Iapetus and its cousins Hyperion and Phoebe show only hints of terrestrial processes. They are truly worlds fantastic.

NOTES
1 Cassini, G. 1705. Quoted in *The Philosophical Transactions of the Royal Society of London Abridged* 1, 368.
2 Morrison, D. et al. 1975. The two faces of Iapetus. *Icarus* 24, 157–171.

# 9 Pluto: The First View of the "Third Zone"

I clearly remember a conversation I had with my brother Bruce when I was six or seven in the small bedroom we shared in our home in the steel town of Bethlehem, Pennsylvania. My brother asked me how long I thought it would take to freeze to death if you were standing on Pluto and weren't wearing a spacesuit. Pluto was one of the nine major planets back then, and as the farthest out, it was the coldest thing a child could imagine. I asked Bruce if he thought Pluto was colder than Antarctica: we both thought it was. We figured if you were bundled up in a snowsuit and covered up with blankets, you might live five minutes at the South Pole. This era was pre-internet – even hand-held calculators were more than a decade away – so we couldn't look up the temperature of Pluto, or even of Antarctica. We finally decided Pluto was so cold you'd die in about five seconds. I remember the view I had of the surface of little Pluto – ice everywhere and very dark. My trusty *Child's Book of the Stars* had only six sentences on Pluto, and its stated size of 4,000 miles was disappointingly small, only about half the size of the Earth. But as we shall see, Pluto shrank even further before its planetary demotion.

Pluto grabbed me again in 1965, at the same time *Mariner 4* sent back its images of a disappointing, cratered, moonlike Mars. Scholastic Book Services, the lifeline of curious children, offered a 45-cent book called *The Search for Planet X* by Tony Simon.[1] Sputnik had been launched eight years earlier, the space race was in full swing, and everyone was looking up. *The Search* was pure romance, offering the improbable story of a farm boy who discovered a planet.

Clyde Tombaugh was drawn to the stars by the wide and clear skies of Kansas. He built his own telescopes, from grinding the lenses to fashioning their mounts out of old parts from farm machinery, including a cream separator. He devoured every astronomy book in the local library. In 1928, when he was 22, he sent a few of his drawings of Mars and Jupiter to Lowell Observatory Director Vesto Slipher, brother of E.C. Slipher and an Indiana farm boy himself who assumed the directorship in 1926. (Lowell died in 1916.) Sensing an affinity with Tombaugh, Slipher was so impressed that he offered Clyde a job. Tombaugh accepted and started in January, 1929.

Percival Lowell's foremost interest was Mars, but he also thought there was a "Planet X" beyond the orbit of Neptune. He based his belief on observations of perturbations, or irregularities, in the orbits of the known planets. The French astronomer Urbain Le Verrier and the British mathematician John Couch Adams had similarly, and independently, predicted the existence and position of Neptune based on its perturbations of Uranus. Their predictions soon led to the discovery of that planet in 1846. Lowell had also done some calculations and predicted the position of a ninth planet. Slipher was too overwhelmed with his other projects – such as measuring the distances to galaxies – to search for Planet X, so he assigned the job to Tombaugh. The pay was almost nil, and the living and working conditions were spartan. His living quarters consisted of a small, unheated room on the second floor of the administrative building. The observatory was cold in the winter and remote, surrounded by a forest filled with wild animals. During a dark observing run, another astronomer had petted what he thought was a dog, but daylight revealed the tracks of a mountain lion.

Working 12- to 14-hour days, Tombaugh spent every clear night taking photographic glass plates with the Lowell Observatory's 13-inch telescope, which could "see" objects more than ten thousand times dimmer than could be seen with the unaided human eye. He concentrated on the constellation Gemini, where Lowell predicted the new planet should be. During daylight hours, he carefully scanned one plate after the other on an instrument called a "blink comparator" (Figure 9.1); the device can still be seen at the Observatory's museum. A moving target such as a planet would appear to flit back and forth

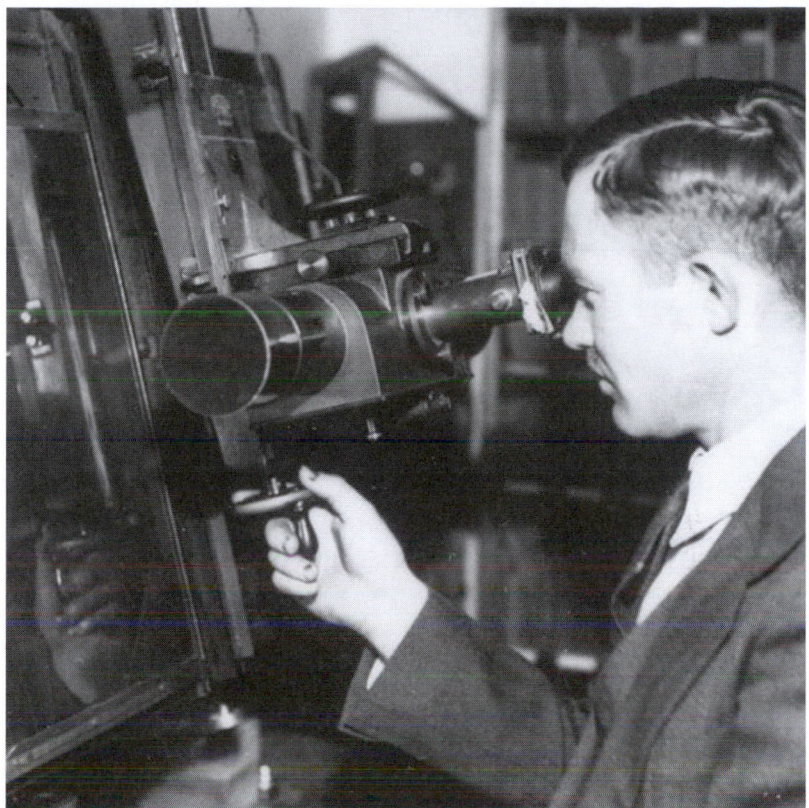

FIGURE 9.1 Clyde Tombaugh with the Lowell Observatory blink comparator he used to discover Pluto in 1930. Courtesy Lowell Observatory Archives.

against the background of fixed stars. After a year of tediously observing millions of stars, Tombaugh finally found something. Photographs obtained on January 21, 23, and 29, 1930 (Figure 9.2) showed a faint object – more than a thousand times fainter than anything that can be seen by the naked eye – moving at about the right velocity for an object at Planet X's distance (more distant objects move more slowly: asteroids, for example, skipped across the plates). On February 18, Tombaugh walked into Slipher's office and declared "Dr. Slipher, I have found your Planet X." An additional confirmation image was taken on February 19, and there was the little planet right where it should be.

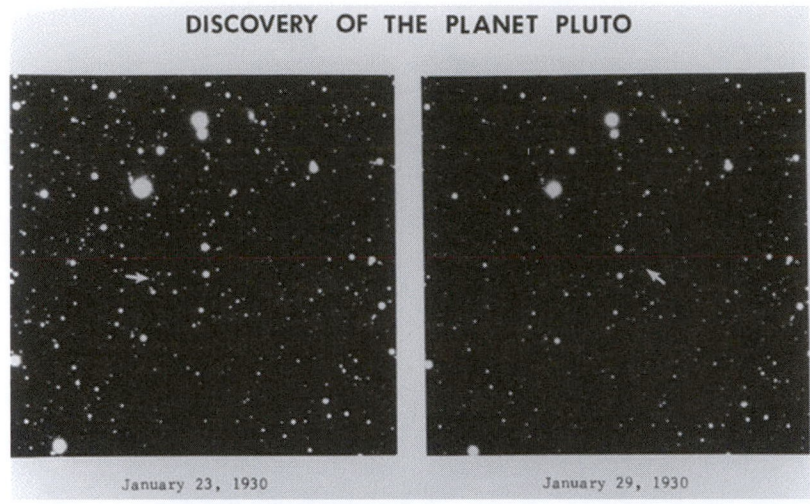

FIGURE 9.2 Two of the images that led to the discovery of Pluto. Against the background of fixed stars, Pluto is a faint, moving dot. Courtesy Lowell Observatory Archives.

The discovery was announced on March 13, 1930, a welcome diversion from the encroaching Great Depression. In an article in the upper left hand corner of its first page, the *New York Times* declared "Ninth planet discovered on edge of Solar System; First found in 84 years" and "Possibly larger than Jupiter and 4,000,000,000 miles away." Quoting C. O. Lampland, the assistant director of the Observatory, the paper said "All observations indicate the object to be the one which Lowell saw mathematically." Tombaugh, identified as a "photographer at the observatory," is credited with first seeing Pluto, but the article quotes Roger Lowell Putnam, Percival Lowell's nephew and the chairman of Lowell Observatory's board of trustees, as stating the "discovery of the new planet may safely be credited to Dr. V. M. Slipher." As the circumstances of the discovery became better known, Tombaugh was fully credited.

Venetia Burney, the 11-year-old daughter of a chaired theology professor at Oxford University, first suggested the name Pluto, the Roman god of the underworld. (The Disney cartoon character came later, which is lucky considering Disney's well-known protection of its copyrights.) The name conveniently started with Percival Lowell's initials. In 2006, still during Burney's lifetime, the dust detector on *New Horizons*, the spacecraft mission that encountered Pluto in 2015, was named the Venetia Burney Student Dust Counter. This instrument was built and managed by students at the University of Colorado.

An image of Pluto taken on the Lowell Observatory's 24-inch, its largest telescope, failed to show a planet-like disk. This finding was disconcerting, and Pluto's discoverers should have realized that its star-like countenance implied it was very small, smaller than the Earth. The pull of Lowell's predictions, which postulated a large planet, coupled with the bandwagon effect and Lowell's Brahmin status, clouded their beliefs about Pluto. But many scientists were skeptical of claims about Pluto from the beginning. Armin Otto Leuschner, a professor of astronomy at the University of California at Berkeley, stated as early as 1932 that Pluto's mass was less than the Earth's and that the discovery had nothing to do with Lowell's predictions. In 1934, the great astronomer William Baade published a paper that stated Pluto would be the same size as Triton, the moon of Neptune, because it was as bright as Triton and about the same distance from the Sun. Baade still couldn't say how large Pluto was, because he didn't know how reflective the surface is: a very dark object can be larger than a small highly reflective body and still appear as bright at a distance. What Baade was saying is that Pluto is small, more moon sized than planet sized. Now we know that Pluto is even smaller than the Earth's Moon. In 1973, my MIT professor Irwin Shapiro told my planetary physics class that he studied Lowell's notebooks and found them to be nonsense. We now know that Tombaugh discovered Pluto with his own hard work rather than from Lowell's predictions.

Tombaugh continued his planet search for the next thirteen years; he scanned the plane of the Solar System for objects nearly ten times a faint as Pluto. He found many asteroids and a comet, but no additional planets.

Reading *The Search for Planet X*, I wondered if my own life could take the improbable turn that Tombaugh's had. I was still worried about those echelons of buzz-cut engineers at Mission Control, not a woman in sight. I remember seeing Dr. Anne MacGregor (played by Lee Meriwether) in the 1966–67 TV show *The Time Tunnel*, but she never played the daring, maverick roles of her male counterparts. But Asimov's main brain in *I Robot* and his other robot stories was computer scientist Susan Calvin. Would overcoming the barrier of my gender be even harder than putting aside Tombaugh's rural upbringing and lack of a formal education? (He later earned bachelor's and master's degrees in astronomy from the University of Kansas.)

I was thrilled to have met Clyde Tombaugh in 1985, when he came to JPL for a workshop on Pluto organized by my colleague Ed Tedesco. Thirty years later I met his children at a *New Horizons* science team meeting. Neither followed in Clyde's footsteps: Alden Tombaugh was a banker and Annette Tombaugh Sitze was an elementary school teacher, but they still had an unmistakable resemblance to their father, with that same twinkling sense of humor and forthrightness. If you're a baseball fan, you'll be pleased to know that Clyde's grandnephew is L.A. Dodgers All-Star pitcher Clayton Kershaw.

Using the clever tools developed over the years, astronomers had a vague outline of what tiny Pluto looked like. By the early 1950s observers had measured the light curve of Pluto, which showed it had bright and dark spots. Exotic ices had been detected on its surface – methane, nitrogen, and carbon monoxide – and Lowell Observatory astronomer Will Grundy showed that the brightest regions of Pluto were enriched in carbon monoxide and nitrogen. Moreover, the abundances of these ices seemed to be changing. Pluto almost certainly includes large amounts of water ice, but at its temperature of −380°F, water ice behaves like a rock, and it seemed to be buried under other more volatile ices that had gone through countless seasonal cycles of sublimation and refreezing.

Pluto was a few pixels in the most powerful camera of the *Hubble Space Telescope*, enough to enable astronomers Alan Stern of

Southwest Research Institute, Marc Buie at Lowell, and Larry Trafton of the University of Texas to construct a blurry map showing bright and dark terrains. Astronomers were also able to get a better handle on the size of Pluto by waiting for a stellar occultation, that rare and prized event when Pluto moves in front of a background star and the planet's size can be directly computed from the time the star is invisible. Now astronomers were certain the diameter of Pluto was 1,466 miles, plus or minus about 10 miles (the *New Horizons* value is 1,475 miles). Occultations also enabled us to measure the density of Pluto's atmosphere, as a star dims only a little after it goes behind a thin atmosphere. Pluto's atmosphere increased between 1989 and 2015, as models predicted, because it was closer to the Sun and the frost on its surface sublimated to bulk up the atmosphere. But still, Pluto's atmospheric pressure is only a few millionths that of the Earth.

In 1978, James Christy discovered Pluto's moon Charon, named after the boatman who carried souls across the river Styx; it was also close to his wife's name Charlene. This naming history led most astronomers to mispronounce Charon as Sharon. There is no "sh" sound in Greek, so we sticklers correctly pronounce the chi in Charon. The moon has a diameter about half that of Pluto, comprising an almost double planet, and it orbits around Pluto every 6.4 days. No other moon is as large in relation to its primary: Charon is 12% the mass of Pluto, while our own Moon is only 1.2% the mass of Earth. Observing the mutual dance of Pluto and Charon, in which the two objects eternally face each other in the end state of tidal evolution, dynamicists refined the mass of Pluto: it is only 0.2% that of the Earth.

Soon after Charon's discovery, astronomers realized that between 1985 and 1990, Charon and Pluto would experience a series of events in which the objects would pass in front of each other. These occurrences would be opportunities to refine our knowledge of the sizes of Pluto and Charon, and to more accurately map bright and dark patches on them, as each body would cover specific areas of the other during their passages. In January, 1985, Ed Tedesco and I were the first to observe one of these mutual events, on the 60-inch tele-scope at Palomar Mountain. We didn't announce it, though, because in a highly correlated Murphy's Law occurrence ("if something can go wrong, it will go wrong"), we had forgotten to refill the container that kept the telescope's CCD detector cold with liquid nitrogen. The event was also off by about an hour, so we weren't completely sure the "occultation" wasn't just the observation of the CCD heating up. Richard Binzel of MIT observed the second event the next month, and it agreed with our timing, so we all published a paper in *Science* describing the first two detections. These events still left us with a blurry map of Pluto, but we refined the size of Charon to 750 miles, plus or minus a mile or two. Charon's size was much easier to deter-mine than Pluto's because it doesn't have an atmosphere.

This discrepancy between our observations and predictions of the first mutual event underscored one essential thing we didn't know as well for Pluto as for the other planets: where it was. Pluto has not yet made one of its 248-year orbits about the Sun since its discovery. The determination of orbits is all about how well you can determine the position of an object over a long time. If only an "arc" of an orbit is determined – as is the case with Pluto – there will be uncertainties in the shape of the arc that will be propagated into the future. Think of a runner you have observed for a short period of time. Say you've estimated she is moving at between seven and eight miles an hour. After observing her for a minute, you'd have a good idea where she is. If you had to predict her position in one hour she could be between seven and eight miles away. Errors in orbit determination propagate in the same way. We did know some basics about Pluto's orbit: it was elliptical, ranging from 30 to 50 astronomical units (AU) from the Sun; it was inclined 15 degrees from the plane of the other planets; and it passed inside Neptune for about 20 years (the last passage was 1979 to 1999). During the early days of planning for the *New Horizons* mission, we were so uncertain of Pluto's position at the time of encounter that we had to plan to take a lot of images of the dark sky, just to make sure we "hit" Pluto. Heroic efforts by the team,

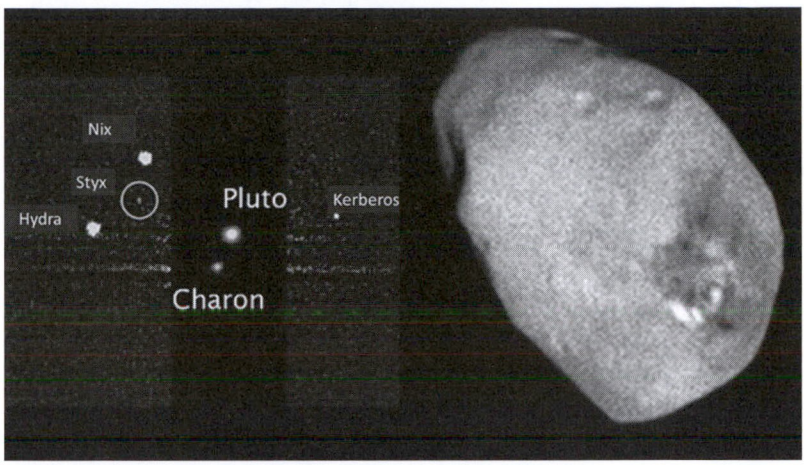

FIGURE 9.3 (Left) Pluto and its moons imaged with the *Hubble Space Telescope.* (Right) An image of Nix returned by the *New Horizons* spacecraft. The longest dimensions of the moons, determined by *New Horizons,* are – Charon: 737 miles; Styx: 4 miles; Nix: 30 miles; Kerberos: 7 miles; Hydra: 34 miles. Pluto has a diameter of 1,475 miles. Like Hyperion, the four smallest moons are in chaotic rotation. NASA/Space Telescope Science Institute/Johns Hopkins University Applied Physics Laboratory. See plate section for color version.

which included digging up old observations and reanalyzing them with new techniques, allowed us to determine the position of Pluto so that we didn't have to waste data on large swaths of dark sky.

Teams working with the *Hubble Space Telescope* and led by Hal Weaver of Johns Hopkins University Applied Physics Laboratory and Mark Showalter of the SETI Institute discovered the other four known moons of Pluto: Hydra, Nix, Kerberos, and Styx (Figure 9.3). They are small, similar to the outer moons of the giant planets, and they are all dynamically related. We believe Pluto's family of moons formed from the same impact, in a scenario similar to our own Moon's formation. In the model developed by planetary dynamicist Robin Canup of Southwest Research Institute, a large impacting body (in the case of the Earth, it was Mars sized) impacted Pluto to send up a great apron of debris, which accreted in a matter of days to form Charon, or our own Moon. The four minor moons of Pluto were the fragments that never accreted. *New Horizons* showed that these small, jagged moons are different colors, signifying different compositions (Figure 9.3). This diversity means the impacting body was likely differentiated into a core, mantle, and crust like the Earth and other terrestrial planets. Like Hyperion, the four smallest known moons are in chaotic rotation.

I had been studying Pluto for years as a tiny point of light, the spirit of Clyde Tombaugh peering back at me from each of the thousands of telescopic images I studied. An army of students at JPL was faithfully helping me to observe Pluto throughout the years to see if the planet had seasons like the Earth. The seasonal cycle of winter and summer on our planet is caused by the tilt of its pole, which causes sunlight to fall at different angles at a specific point on the surface throughout the year: slanted angles mean lower temperatures. Pluto has a similar, but even more extreme, tilt. Intensifying the seasonal effects is the highly elongated shape of Pluto's orbit. Planetary astronomers led by Jim Elliot of MIT, and Leslie Young and Cathy Olkin of Southwest Research Institute, had watched the extraordinarily thin atmosphere of Pluto get denser after its closest passage to the Sun in 1989, as if frost had been removed from its surface and ended up in its atmosphere. My students and I were measuring Pluto's light curve (Chapter 8) to see if we could detect any change in frost patterns on Pluto. This frost is not the water ice so familiar to inhabitants of Earth – it is so cold on Pluto that water behaves like a rock and never melts or sublimates at the surface – but nitrogen and possibly some carbon monoxide or methane. The gas that makes up about 80% of our atmosphere is the frost that appears to crawl over the surface of Pluto.

Most people have a completely inaccurate view of what happens during an observing run. They picture the classic scene of a man in a suit looking through the eyepiece of a large telescope. Decades have passed since observing has been anything like this picture: telescopes are now hooked up to sophisticated charge-coupled device (CCD) imaging cameras – similar to the one in your cell phone but much more sensitive – and spectrometers. The current observers' room looks like a hi-tech surgery theater, with computers and the latest imaging technology. The stress and sense of timing is also similar to that of a surgical procedure: many celestial events such as occultations and transits – when one object moves behind or in front of another – happen at precise times, when all the equipment must be running smoothly and pointed to a cloudless sky. Tombaugh suffered long, cold nights while he searched for Pluto, while we astronomers of today suffer long warm nights in a heated control room dealing with the usual frustrations of technology.

But the low-tech aspects that are still there can bring back the romance of observing. My favorite place to observe is Palomar Mountain. Since most telescopes are on high mountains, the journey up is one of suspense, from the first glimpse of the silvery dome to the drive into the parking lot, where that same dome bears down on the smallness of car and driver. Inside the dome, Palomar's 200-inch telescope lumbers and squeals whenever it is moved, and the smell of pump oil saturates the clear mountain air. Often when I am observing and the night assistant places the telescope on a dim object requiring a long exposure time, I walk out of the observer's room, close the door behind me, and enter the darkness of the dome itself. With the stars peering down the open slit in the dome like so many scattered, immobile fireflies, I simply stare in awe at the same great instrument that saw first light under the supervision of the legendary Edwin Hubble, and with which Maarten Schmidt discovered quasars, the luminous cores of galaxies in their birth pangs at the edge of the Universe.

My students and I noticed that the light curve of Pluto seemed to change around the turn of the millennium. Moreover, it seemed to be changing where Pluto was brightest. The effect was especially evident if observations all the way back to the early 1950s were included in the mix. After all, Pluto's orbit around the Sun took 248 Earth years, so changes such as growing and waning polar caps would take place on decades-long time scales, rather than months, as on the Earth. Along with other scientists, we predicted that frost was moving on its surface, in cadence with the seasons. We also predicted that plumes would appear in the frosty regions, because both Mars and Triton – the other celestial bodies with disappearing polar caps – had plumes or plume deposits.

But we knew so much less about Pluto than the other eight planets that had been explored by spacecraft: it would take a mission to understand Pluto. The two *Voyager* spacecraft that executed the grand tour of the Solar System had successfully reconnoitered the four outer planets between 1979 and 1989. (Pluto wasn't in the right place in its orbit for the *Voyagers* to meet it, although as we stated in Chapter 7, *Voyager 1* could have been redirected toward Pluto, rather than to a close flyby of Titan.) The mission's culmination was the exploration of aquamarine Neptune with its raging white storms, and its large moon Triton, which was discovered in 1846 within a month after Neptune itself was added to the array of known planets.

Triton was saving a secret that became the final surprise of the mission: active plumes on its surface. As *Voyager 2* approached the moon, a giant polar cap with numerous streaks, like deposits from geysers, appeared (Figure 9.4). The imaging team spotted a plume shooting up several miles from the surface and being entrained in Triton's rapidly moving troposphere at the base of its thin atmosphere. The plumes are believed to form in the same way as Mars's intriguing spiders: pockets of gases heated by sunlight build up pressure until an explosive eruption occurs.

Later work by a JPL team using *Hubble*, led by my JPL colleague James Bauer, showed that the polar cap on Triton had continued to sublimate. Like the Earth, Triton has seasons, except they play out over a 165-year period, the time it takes Neptune to orbit the Sun. Echoing Baade's ideas from a half-century ago, many astronomers started to see Triton as Pluto's twin. Triton was likely a captured Kuiper Belt Object (KBO).

The effort to explore Pluto unfolded in a series of fits and starts. Since the close of the *Apollo* era, the crewed part of space exploration – NASA's core mission – was in decline, and the Agency

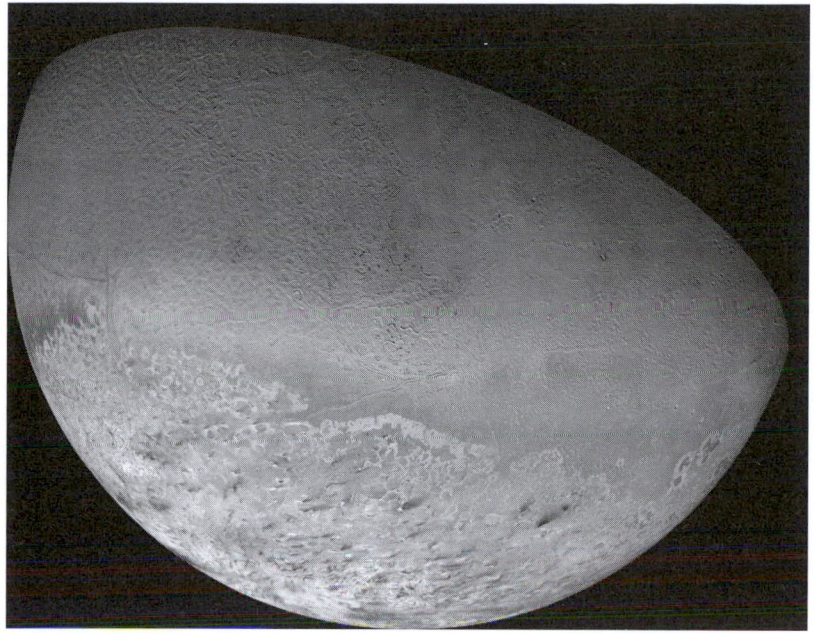

FIGURE 9.4 Triton, the largest moon of Neptune, with a polar cap and plumes. Its diameter is 1,682 miles. NASA/JPL-Caltech.

was in the grip of "faster, better, cheaper" ("pick two," many engineers and scientists were quick to add). In the last decade of the last millennium, a "Pluto Underground" had led the development of the *Pluto Kuiper Express* (*PKE*) to the only unexplored planet. NASA gave the mission high priority, but by 2000 *PKE* had ballooned in cost and was cancelled. Pluto scientists were frantic: not only was Pluto past its closest approach to the Sun, when its transient atmosphere might begin collapsing, but Jupiter was moving away from the position where it could serve as a slingshot to propel the spacecraft to distant Pluto. Through the extraordinary efforts, led by Stamatios "Tom" Krimigis and Alan Stern, NASA agreed to fund a mid-cost (three-quarters of a billion dollars) Pluto mission. The *New Horizons* mission, led by the Johns Hopkins Applied Physics Laboratory (APL), which Krimigis directed at the time, was selected. But it continued to suffer funding threats until the National Academy of Sciences, the nation's premier advisory group, gave Pluto a high priority for exploration.

Another threat came from an unexpected quarter. Los Alamos Laboratory was set to provide the plutonium to power the spacecraft (the Sun is too dim at Pluto to use solar panels), but numerous security lapses at the government laboratory led to its partial closure, and the plutonium could not be delivered. Glen Fountain, the superb *New Horizons* project manager, cut the spacecraft's energy requirements, and enough plutonium was delivered just in time. The spacecraft was finally launched on January 19, 2006, also in the nick of time. Pluto was still a planet, but not for long.

While the battle to get a Pluto mission was being fought, Pluto was fighting for its own life. The movement against Pluto's planet-hood began as a still small voice, but it built into a crescendo beginning in 1992, when David Jewitt and Jane Luu used Hawaii's Mauna Kea Observatory to discover the small body 1992 QB1 beyond the orbit of Neptune. Six months later they discovered another object: it looked like Pluto might be a member of a second asteroid belt.

By the turn of the millennium, dozens of objects beyond Neptune had been discovered. Collectively they are known as Trans-Neptunian Objects or Kuiper Belt Objects, named after Gerard Kuiper, one of the astronomers who suggested their existence. Now it seemed the tiniest planet might suffer the same fate as Ceres, the first and largest asteroid discovered in 1801, which reigned as a planet for decades. Brian Marsden, the head of the Minor Planet Center of the Harvard-Smithsonian Center for Astrophysics, and a powerful force in the International Astronomical Union (IAU), the organization responsible for the status and naming of all things astronomical, was pushing to have Pluto reclassified as a "minor" planet. Even more hurtful was an exhibit at the Hayden Plane-tarium in New York City that showed only eight planets: all the children filing past declared "Where's Pluto?" In this new vision of the Solar System, the icy bodies beyond Neptune populated a "third zone" separate from the four rocky terrestrial planets,

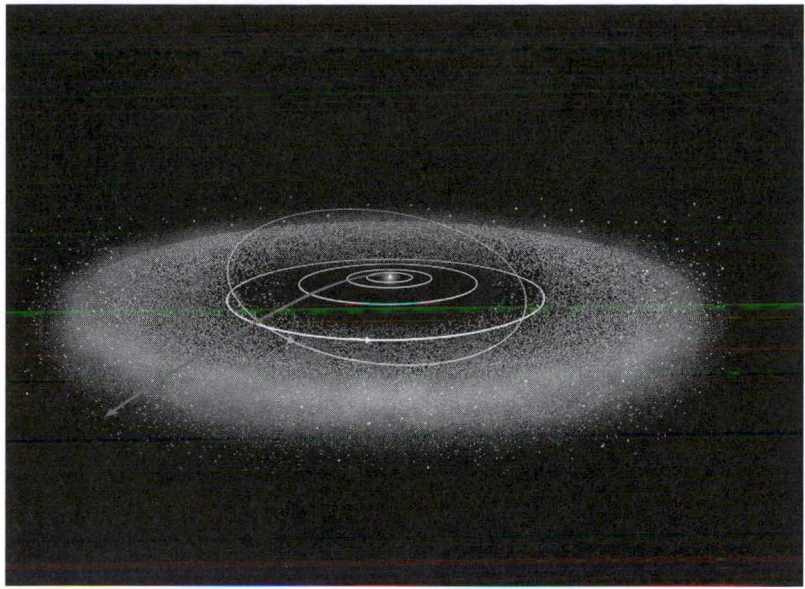

FIGURE 9.5 The Kuiper Belt, showing the orbits of the four outer gaseous planets, Pluto's inclined orbit, and the trajectory of the *New Horizons* spacecraft. NASA/JPL-Caltech.

of which the Earth is a member, and the four gaseous planets (Figure 9.5).

Like the asteroids, these icy bodies are remnants of the Solar System's formation, the building blocks of the main planets. Now we know of over 1,200 KBOs, of about 100,000 that are estimated to exist. The largest group of "cold classical" KBOs has relatively undisturbed orbits between 30 and 55 AU from the Sun, while a second class of KBOs, known as scattered disk objects, have elliptical orbits that are often inclined to the plane of the Solar System and can reach as far as 1,000 AU. These KBOs – which are sometimes not included in the Kuiper Belt at all, but in a separate category of Trans-Neptunian Objects dubbed Sednoids after their largest member Sedna – may have been disturbed in an early migration of the giant planets from the inner to the outer Solar System, an idea known as the "Nice Model," named after the city on the French Riviera where a group of summering astronomers formed this hypothesis. This third zone thus contains the fingerprints of the early dynamical evolution of the Solar System. Like the asteroids, the KBOs differ in composition, offering clues to the physical conditions of the cloud of gas and dust from which the planets formed. In recognition of the Kuiper Belt's importance to understanding the formation, evolution, and structure of the planets, the IAU names its objects after creatures associated with creation mythologies of the world. Many KBOs have moons, and one even has a ring (10199 Chariklo).

The groundswell against Pluto intensified in 2005 when a team led by astronomer Michael Brown of Caltech announced the discovery of 2003 UB313, which was officially named Eris by the IAU, after the goddess of discord. (Brown had earlier nicknamed the new object Xena, and its moon Gabriella, after television's cheesy pop culture icons, the ancient Amazonian warrior princess and her sidekick.)[2] A measurement of thermal infrared emission suggested its size might be larger than Pluto, although many astronomers were skeptical, as the errors on this difficult measurement were large (I was in the skeptics' camp). In August 2006, during a meeting in Prague, after much discussion and several proposals, the IAU elected to demote Pluto to a "dwarf planet," a designation it currently shares with Ceres and three other KBOs: Eris, Makemake (named after the Hawaiian goddess of childbirth), and Haumea (named after the Rapa Nui creator god). Pluto failed one of the criteria the IAU had set for a planet: it had not "cleared its orbit," since there were other KBOs in its vicinity. The other criteria are that it orbit the Sun and that it have enough gravity to be round, both of which Pluto met. Some detected an anti-American bias. Pluto was the only planet discovered by an American, and a European meeting had killed it. I'm a member of the IAU, but I couldn't attend the meeting as I was driving my youngest son to his first year of college at the same time.

Those making the decision to discard Pluto weren't primarily planetary scientists, and the requirement of clearing an orbit was a dynamical concept – which would be foremost in an astrophysicist's way of thinking rather than that of a planetary scientist, who would consider the processes occurring on its surface. We planetary scientists often catch ourselves calling Titan a planet, or even Io (of course this designation is incorrect, as neither object orbits the Sun). "Clearing its orbit" is also vague – has the Earth not cleared its orbit because there are many NEOs in its vicinity, or because it hasn't gobbled up the other terrestrial planets? As my colleague and planetary dynamicist Mark Showalter points out: if you were an alien entering our Solar System, you would see four planets: the gas giants (1,000 Earths can fit inside Jupiter). All other objects would be "dwarf," with the four inner terrestrial planets forming an inner zone of minor objects. But, as Figure 9.5 shows, the Kuiper Belt is a massive collection of objects similar to the asteroid belt, and singling out Pluto for special designation is scientifically suspect.

In the end, these definitions are not very important, and anyway Pluto is still classified as a dwarf planet. The questions we astronomers were asking about Pluto were so much more interesting: Is it geologically active? How stable is its atmosphere? Does it have seasons like the Earth? Does it have liquid water under its surface? Along with many other planetary scientists, I still refer to Pluto as a planet and include it in any discussion of the planets.

Perhaps the saddest, sorriest part of Pluto's dethronement was the disappointment felt by the children of the world, who had their stories taken away, the story of the farm boy from Kansas who discovered a planet, and the story of Venetia Burney who named it. Clyde and Venetia were real, more real than Podkayne of Mars, or Ray Bradbury's flawed humans in the *Martian Chronicles*. These heroes, and the childhood queries such as I had with my brother, are what move children to become scientists, or just critical thinkers. As a planetary astronomer I was annoyed that the public's interest in Pluto centered mainly around the question of its planethood. (My most humorous incident occurred in 2005, when I suffered a serious bout of Lyme disease. My doctor was about to perform a spinal tap, and he blurted out: Is Pluto a Planet? If his intent was to draw my mind away from the unpleasant procedure, it worked. I remember talking about Pluto for the whole time, but I remember nothing of the procedure itself.)

In January, 2016 the world was startled by the announcement that there might really be a ninth planet: a body larger than the Earth moving in an orbit as close as 20 billion miles from the Sun and as far as 100 billion miles. Michael Brown and Konstantin Batygin, also at Caltech, based their claim on the similar orbital alignments and tilts of the six known sednoids: an invisible gravitational tug by the new planet had nudged them into this configuration. Superb intuition backed up by solid numerical calculations had hatched a daring and provocative idea ripe for testing. Brown said he should be able to find the planet in five years.[3] Others were rightly skeptical. Maybe those six worlds are only a small, biased sample of what is out there. Other scientists fled to their telescopes and high-speed computers to look, calculate, and think. Again, the edges of science were marked by doubt and controversy, pushing those edges out even farther.

Planetary encounters are occasions of high excitement among scientists. Not only are our predictions confirmed or demolished, but we are gradually approaching something we have never seen. The preparation takes years as every team involved in the flyby plans each and every detail and tests it over and over. Scientists decide where to go and how close we should approach our targets, and mission planners design trajectories that will get us there. Then we plan our observations, with all the instruments efficiently working together, each pushing to take their highest priority measurement within the constraints of the small amount of data and power available. The science data is stored on a computer and sent from the spacecraft's large radio antenna to the NASA's Deep Space Network (DSN) of radio telescopes in Madrid, Canberra, and Goldstone, in California's Mojave Dessert. Pluto is so far away, that the data is sent back in a trickle compared to the flood of a lunar or even a Mars mission. The *New Horizons* engineering and science teams even rehearsed the encounter, walking the spacecraft through the commands it would execute in the hours around the encounter. You can't go out and fix the spacecraft if it doesn't work correctly, so we had to make sure everything was robust.

The incessant drumbeat of trajectory correction maneuvers, calibrations, software uploads, tests, "rehearsals," and more finally yields to the lyrical tale that each unexplored world bestows upon those who await its first appearance. Alan Stern, the charismatic and disciplined leader of the *New Horizons* team, had his entire science team resident at the Applied Physics Lab (APL) in Maryland for two weeks.

The encounter with Pluto came exactly 50 years after the *Mariner 4* spacecraft flew past the Red Planet. On July 14, 1965, I was plastered to my parents' TV screen as those first disappointing images of a cratered desolate Mars came in. A half-century later, I was a full-fledged astronomer on the inside of the *New Horizons* mission as a science team member, awaiting those first images from the planet that is only second to Mars in holding the public's imagination. And Stern's team was filled with women, including the Mission Manager Alice Bowman, the three Deputy Project Scientists, several science team members, and more junior recent PhDs. Perhaps the best part was we had 5,000 times more images than *Mariner 4*.

But then, a final challenge presented itself only ten days before the encounter. I received this terrible message from Stern on the 4th of July: "This afternoon at approximately 1:54 pm EDT, we lost contact with *New Horizons*. Later, about 3:15 pm EDT, communications were established after the spacecraft went to safe mode . . . We are no longer collecting science . . . The cause of the safing event is thought to be a work overload on the C&DH 1 processor due to a combination of the flash burn of the CORE load simultaneous with heavy CPU cycles compressing forward data." As a JPL veteran, I knew a spacecraft went into safe mode only when a serious problem occurred, and this error sounded very bad.

Stern and Glen Fountain summoned Mission Manager Alice Bowman and her team to APL and they worked over the 4th of July weekend to fix *New Horizons*. The team programmed a series of commands to restore the functioning of the onboard computer and sent it up to the spacecraft. It worked. But there was a loss: while all this engineering moxie was going on, the most sensitive observation to search for more moons of Pluto was pushed aside.

Our embrace of Pluto resumed, and the blurry features on Pluto gave way to wild speculations. There was a dark feature on Pluto – nicknamed "the Whale" that had seemed to be comprised of five or six wavy areas and even had a circular "blowhole" at one end (it turned out to be a crater). Charon had a dark pole, which several team members soon speculated was deposited in a way similar to the dust that rained down in Iapetus. I was busy looking for plumes.

Since we would rather be taking data than wasting time sending it back to Earth – the spacecraft can't point at Pluto and the Earth at the same time – we decided to send back only a handful of data during the encounter and wait until November, 2015 to get the bulk of it via the DSN. We got one image of Pluto's full disk just prior to encounter, and three images from a mosaic of its surface at closest approach, and five images from that same mosaic five days later. And even these images were compressed, leaving out many details. This small subset of our cache revealed an exotic kingdom of ice we could have never imagined.

Figure 9.6 shows one hemisphere of Pluto: it's the famous picture with the broken heart. There is at least one polar cap on Pluto and possibly two, but the brightest area is the "heart," which the *New Horizons* Team named Tombaugh Regio (TR). It is as bright as one or two-day old snow in New York City. That means of course it is fresh. It is also devoid of craters, another sign it was formed by some recent activity. A closer look shows that this bright region appears to be composed of an array of glaciers, each flowing into each other to form an overlapping fish-scale pattern (Figure 9.7). These are the first confirmed glaciers outside the Earth (Titan and Mars may also

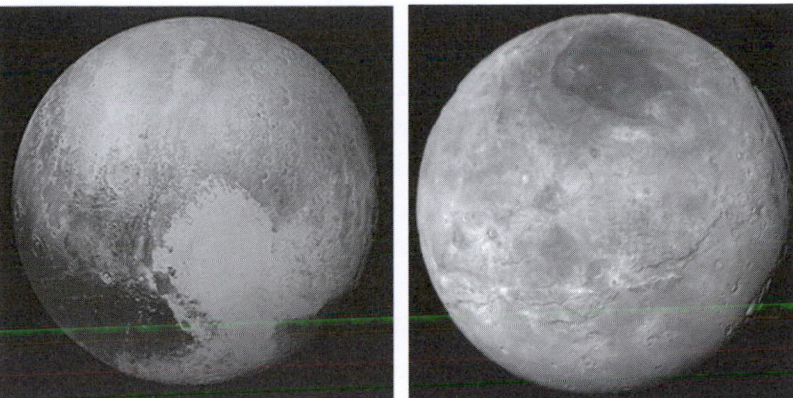

FIGURE 9.6 On the left side is the last full-disk image of Pluto returned before encounter, showing the diversity of features on its surface, including at least one polar cap, and very dark and bright regions. The brightest regions are crater free, implying a recent formation, while craters appear in the older darker terrain. On the right is Charon, with a dark polar cap and system of faults. Charon is about half the size of Pluto. NASA/Johns Hopkins Applied Physics Laboratory/Southwest Research Institute.

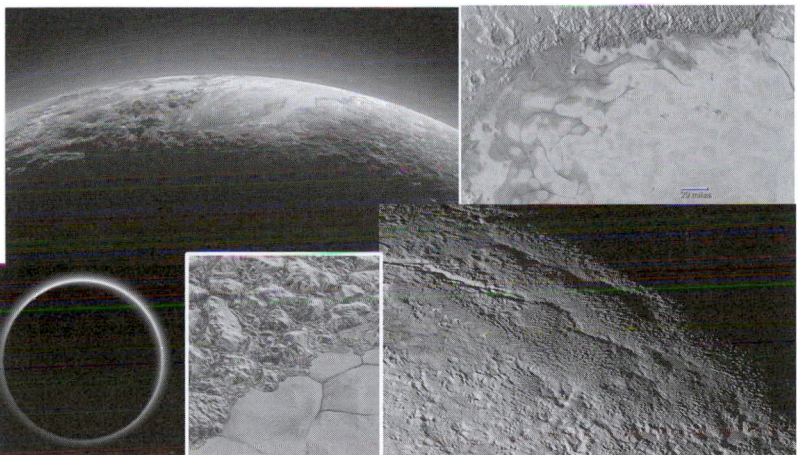

FIGURE 9.7 Images from Pluto. Clockwise from upper left: a glacier with a possible extinct ice volcano to its left and atmospheric haze; an array of nitrogen glaciers on Pluto; the enigmatic "snakeskin terrain," of unknown origin; the interface between Pluto's glaciers and surrounding mountains; an image taken after closest approach revealing the back-lit haze layer in Pluto's upper atmosphere. NASA/Johns Hopkins University Applied Physics Laboratory/Southwest Research Institute.

harbor them), but they are nitrogen glaciers, with an icing of carbon monoxide and methane frost. We still don't know what caused TR to form. Was there an eruption from a subsurface ocean? Did an impacting object punch through Pluto's crust all the way to this ocean? Why does it lie directly opposite from the side of Pluto that faces Charon? Is there an energy imbalance on Pluto? Tombaugh Regio was the region that *Hubble* and ground-based observations showed was sublimating away.

Other images showed mountains, including spire-like structures between the individual glaciers. The separate regions of the "whale" dissipated as we got closer, but this dark region had craters, suggesting it was Pluto's oldest terrain. Pluto had a greater range of reflectivity than any other body in the Solar System, except for Iapetus. I thought I saw plume deposits, but my colleagues were skeptical. One of the most astonishing views was a parting shot of Pluto backlit by the Sun. This angle can never be seen from the Earth, as Pluto is always fully illuminated from our vantage point. Pluto had a spectacular aura of haze surrounding its globe. Haze is caused by dust, and we were soon speculating that Pluto had another Earthlike process that is also seen on Titan: the formation by the bombardment of high-energy ultraviolet light of complex organic molecules in its upper atmosphere This haze – similar to Earth's smog – falls to the surface of Pluto to impart its reddish hue. We saw the same process on Titan.

Charon didn't disappoint either (see Figure 9.6). We kept saying that Charon – which we expected to be a "dead" object – was about as exciting as Pluto was supposed to be. There were faults on its surface, implying an active geological period, and the dark pole appeared at first to be a giant impact basin that opened up its crust to reveal the mantle underneath. Further scrutiny and modeling by Will Grundy of Lowell Observatory and his team suggested the dark reddish pole is instead a chemical change induced by methane traveling to the moon from Pluto's atmosphere. The small moons were battered, irregular objects, as expected.

And the journey of *New Horizons* does not end here. NASA approved an extended mission to visit the KBO 2014 MU69 on New Year's Day 2019. This KBO is only tens of kilometers in size, and it lies another billion miles from the Sun. An investigation of a smaller KBO, more representative of the tens of thousands of objects in the Kuiper Belt, will be a good complement to our close study of Pluto, the largest known KBO.

In the half-century between *Mariner IV* and *New Horizons*, I learned that science is not really about computing or studying hard in your math and science courses (although all those things are necessary to its success). It is about inspiration, followed by persistence. Under the shadows of the mighty blast furnaces of Bethlehem Steel, where my Dad labored every day, a young girl could dream and think. What my early life taught me is that non-conformity can come from within. I didn't let those legions of buzz-cut engineers scare me away from what I really wanted to do. When I heard Malvina Reynolds' folksong "Little Boxes" as a teenager, I could acknowledge its celebration of distinctiveness, but I could also envision an underground of children standing in the little back yards of those "boxes" looking skyward, or experimenting in their garages, or creating novels and music. Perhaps even their parents, seemingly slaves to middle-class conventions, were writing or looking through telescopes, too. Everyone has an interior life that, to quote Walt Whitman, contains multitudes.

Pluto also possesses much more than we ever envisioned. It is way more than something that isn't a planet. It has more charisma than Neptune or Uranus: in that realm it ranks with Mars and the beautiful Saturn, encircled by a different crown of glory. With the completion of the reconnaissance of the Solar System on July 14, 2015, when *New Horizons* encountered dwarf planet Pluto, we scientists realized that if there was ever a time when the destination and not the journey counted, here it was. Pluto is almost a little Earth.

NOTES

1 Simon, T. 1965. *The Search for Planet X*. New York: Scholastic Book Services.

2 Brown, M. 2012. *How I Killed Pluto and Why It Had It Coming*. New York: Spiegel and Grau.

3 Chang, K. 2016. "There May be a Ninth Planet (Not You, Pluto)." *The New York Times*, January 21, 2016, p. 1.

# 10 Earths Above: The Search for Exoplanets and Life in the Universe

For a few short moments in my life I thought I had something to teach the Teacher of the Dalai Lama. I was in my fourth year of graduate school at Cornell, settling in on my PhD thesis topic to understand what the surface of Europa is like. I was busy analyzing some *Voyager 2* images of the surface of Europa in the Spacecraft Planetary Imaging Facility (SPIF). The Teacher of the Dalai Lama was coming to Cornell to see Carl Sagan, and he wanted a little tour of SPIF. Since I knew a lot about SPIF – I practically lived there at that time – Sagan had tasked me with giving the tour. Enrobed in cascades of maroon and gold, the ethereal monk glided into SPIF with an entourage of intimidating British interpreters, scholars, and disciples. I thought the Teacher would share my excitement about this ice world in space. I told him how smooth the surface was, how there was probably an ocean beneath its cracked icy surface, and – jokingly – that it might be fun to ski on Europa. I had heard that the Dalai Lama had a sense of humor, but his teacher apparently not so much. He listened politely to my lecture, but he looked at me sternly after I had finished, and he said (at least his interpreter said): "There are many worlds in this Universe that you cannot even envision."

The Teacher was of course correct, even in the scientific sense. At that time, not a single planet outside our Solar System had been discovered, but we thought there must be billions. Exoplanets, or extrasolar planets – planets around other stars, and extrasolar systems – planetary systems around other stars – were hypothesized to not only exist but to abound in the Universe. A famous set of images taken by the *Hubble Space Telescope*, the deep field surveys, drives home the immensity of our Universe and the possible number of worlds within it. Figure 10.1 shows the *Hubble* Ultra Deep Field

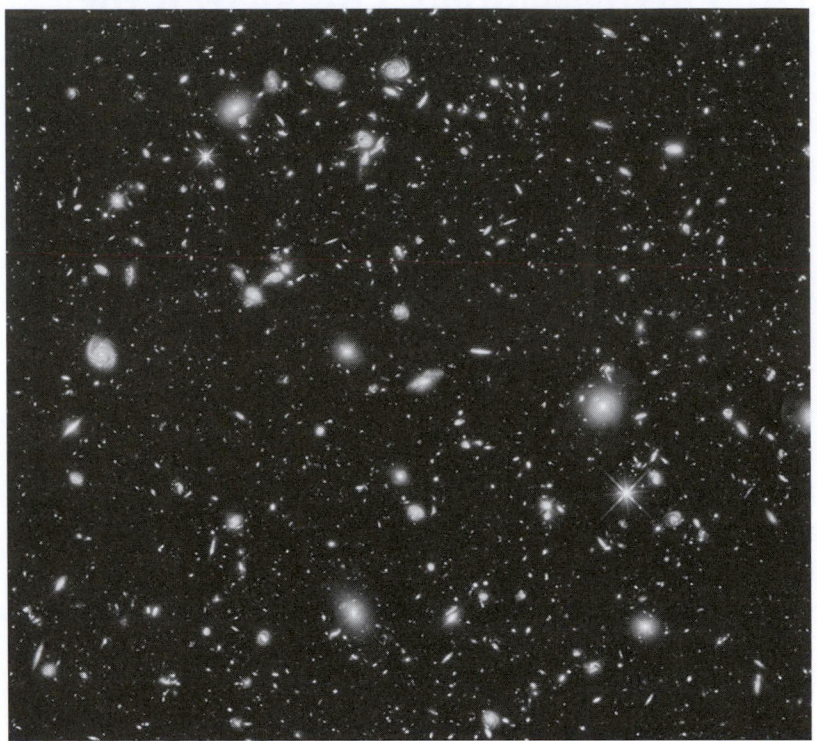

FIGURE 10.1 The *Hubble* Ultra Deep Field survey, constructed from images obtained in 2003 and 2004, and showing about 10,000 galaxies. NASA/ESA. See plate section for color version.

survey, constructed of hundreds of images with a total exposure time of nearly one million seconds and covering one 13-millionths of the sky. Within the image are 10,000 objects, nearly all galaxies, each of which is a collection of typically hundreds of billions of stars and which comprise the main organizational unit of our Universe. Our own Milky Way is a garden-variety spiral galaxy. Multiply 13 million by 10,000, and you get over 100 billion galaxies in the Universe, each of which may have billions of solar systems, with planets, moons, and possibly inhabitants.

The ancient Greeks, in particular the empiricists such as Anaximander and Democritus, proposed the existence of multiple worlds, but these worlds were not associated with individual stars.

Giordano Bruno, the victim of the Roman Inquisition whom we met in Chapter 5, advanced three core ideas of extrasolar systems: that the stars were suns like our own; that these stars had planets; and that these planets could harbor life. Isaac Newton also believed the stars to be suns with their own system of planets. Herschel was certain there was life on the Moon. In Chapter 1 we saw that the idea of the plurality of worlds, including inhabited planets, was widely accepted among nineteenth-century scientists.

Astronomers began looking for exoplanets almost as soon as it was possible to do so, as they were always regarded as one of the holy grails of astronomy. The earliest efforts used a basic astronomical technique known as astrometry: measuring the position of a star or planet very precisely on a photographic plate (now CCD detector). If a star had a large planetary companion, the planet would tug the star back and forth as it orbited. This tugging should be detectable as a wobble on a photographic plate that could be measured by standard astrometric techniques. As early as 1855, several astronomers began to claim that the nearby (16.6 light years) binary star 70 Ophiuchi had a planetary companion, which led to its being established as a favorite locale for science fiction, including the *Dune* series and the eighth episode of the original *Star Trek* series, in which Kirk, Spock, Bones, and some assistants find themselves on a terrestrial twin. Eventually a series of careful observations discredited the existence of a planet around 70 Ophiuchi. In another long-standing example, Peter van de Kamp of Swarthmore College's Sproul Observatory claimed, in the 1960s, that he saw a wobble in the position of Barnard's star, which he explained by the presence of a Jupiter-sized planet (later two planets). Many astronomers accepted his claim at first, but in the way that science is self-correcting, a series of careful observations was unable to confirm it. Worse, it seemed the wobbling in the star's position was correlated with the telescope's schedule of having its lenses cleaned. Van de Kamp was measuring a systematic instrumental effect rather than the pull of an exoplanet! These and other early false detections using astrometric techniques had given the search for exoplanets a checkered reputation that has since been overcome. Only a handful of exoplanets have been discovered by astrometric search techniques.

The first exoplanets were confirmed in the early 1990s, and one was even orbiting a pulsar – the dense, rapidly rotating core of a dead star. But the detection in 1995 of a planet around 51 Pegasi, a Sun-like star just barely visible to the unaided eye, was a watershed event. After some initial skepticism that the detection may have represented a pulsating stellar atmosphere rather than a giant planet (I remember a heated, but collegial, debate between advocates of both sides at an annual meeting of the Division for Planetary Sciences), the radial velocity technique used to discover the star's planetary companion was widely used by a number of groups to detect additional planets. By the turn of the millennium 30 exoplanets were confirmed.

The radial velocity, or Doppler, technique depends on the precise measurement of the position of lines in a star's spectrum. A star's radial velocity is its speed with respect to Earth. As a planet orbits a star, it literally pulls it to and fro. When the planet pulls the star away from the Earth's line of sight, its light signal is red-shifted; when the star is pulled toward the observer it is blue-shifted. Recall that red-shift is the Doppler effect: as a body recedes from us, its wavelength is increased, or moved toward the red end of the spectrum. One simple way to conceptualize this effect is to think of the waveforms as being "stretched out." A corresponding blue-shift occurs when the planet tugs the star toward us. A planetary companion thus causes its star to exhibit a regular pattern of blue-shifted and red-shifted spectral lines. Astronomers choose a line in the star's spectrum, a fingerprint of one of the many elements that are found in stellar atmospheres, and compare it to a laboratory sample at rest. The technique is simple in principle, but difficult in practice, because the measured shifts are very small. Geoffrey Marcy of San Francisco State University and the University of California at Berkeley and his colleagues pioneered the use of this technique.

The early discoveries described large planets orbiting close to their companion stars – so-called "hot Jupiters," that were unlike

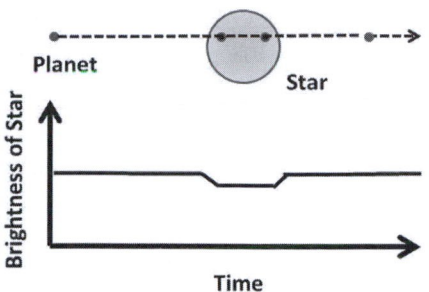

FIGURE 10.2 A diagram of the brightness of a star as a planet transits in front of it.

anything in our Solar System. But large planets should be the first ones to be discovered because they are the easiest to detect. The more massive an object is, the larger the tug it exerts on its central star. Planets near their stars are also easier to detect, as we can measure the complete cycle of blue- and red-shifts quickly. We would have to wait 12 years for Jupiter to exhibit its full cycle of the Doppler effect on the Sun's radial velocity. Astronomers were somewhat disappointed that the first exoplanets to be detected were so unlike any of our own planetary family.

The transit technique is another bread-and-butter trick in the astronomer's repertoire. As a planet moves in front of a star (the "transit," the same thing as transits of Mercury and Venus discussed in Chapters 1 and 2) its dark night-time disk occults a fraction of the star's light. It is a small fraction, proportional to the fraction of projected area of the planet and its sun, typically about one percent or so, with a correspondingly small decrease in the measured light from the star (Figure 10.2). Again, the technique is simple but difficult in practice, because the measured decrease is so small. The first such detection occurred in 1999, but it was space-based detection of thousands of such transits that led to the next watershed in the quest to find and characterize exoplanets. The transit technique also favors close-in planets, simply because their transits occur more often, as well as larger planets that cause a bigger dip in brightness. The plane of the transiting planet needs to be closely aligned with our line-of-sight; otherwise no transit will occur. The probability of a favorable alignment depends on the size of the planet and its distance from the star: these considerations lead to close-in large planets being even further favored.

There are other more exotic techniques for discovering planets around other stars, including gravitational lensing, in which planets cause anomalies in the magnification expected of massive stars acting as giant cosmic lenses, or the detection of the subtle gravitational effects of one exoplanet upon another. Starting in 2008, planets have been discovered by directly imaging them. The tiny lights of a sun's planetary system are gobbled up by the overwhelming brightness of the parent star. For example, our Sun is 9 billion times brighter than Jupiter and 10 billion times as bright as the Earth. If you were an alien standing on a planet orbiting the nearest star (Proxima Centauri, 2.4 light-years away), imaging the Earth would be akin to taking a photo from Hawaii of a firefly 100 feet away from a thousand bright search lights amassed in the middle of New York's Madison Square Garden.

But astronomers have developed another standard tool – the occulting disk – to block out the star's glaring halo. When combined with adaptive optics – a collection of lenses that compensate for distortions in telescope images caused by Earth's pulsating atmosphere – this technique has detected a few dozen planets. They are large bright planets far from the central star, akin to our own collection of gaseous planets. Imaging thus provides a method of detection that is complementary to the radial velocity and transit techniques. Astronomers have been able to do some rudimentary characterization of exoplanets from direct imaging, such as calculate their mass from the amount of heat they emit, or detect molecules such as methane. Figure 10.3 shows an image obtained with the Keck telescope of the four known planets of HR 8799, which are seven to ten times the mass of Jupiter and orbit 14.5 to 68 AU from the primary.

But it was the *Kepler* spacecraft's automated, mass-based detection of thousands of possible transits that moved the study of exoplanets from detection to characterization. Launched in 2009, *Kepler*

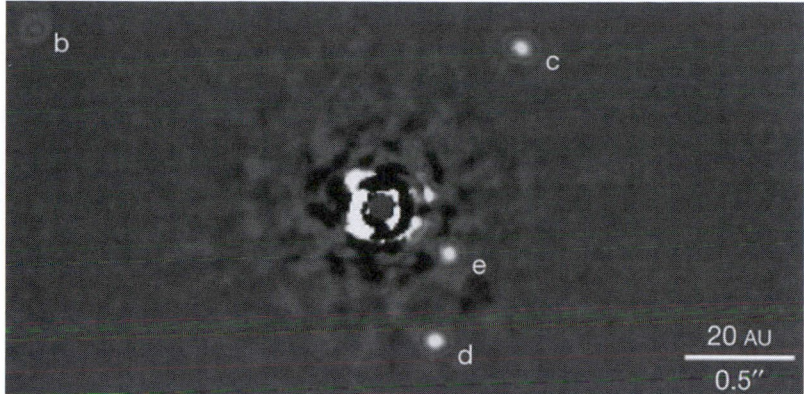

FIGURE 10.3 A direct image of the four planets of HR 8799 obtained on the Keck telescope. 0.5 inches is half an arcsecond, where there are 3,600 arcseconds in one degree of sky. Image by Christian Marois of the National Research Council Canada, Herzberg Institute of Astrophysics and his colleagues. Courtesy University of California. See plate section for color version.

was designed as a low-cost ($600 million) discovery mission with the primary science goal of finding Earthlike planets that might support life. Its single instrument is a camera capable of imaging a region of the sky equal to 12 × 12 degrees: this area is nearly 600 times larger than the apparent size of the Moon and a few thousand times larger than the area captured by a typical telescope. The camera seeks transits by monitoring the brightness of over 150,000 stable stars in this large field of view simultaneously. With so many stars watched closely, thousands of transits will occur and scientists can then estimate from this number the true number of exoplanets of a given size and orbital distance. *Kepler*'s gift is the power of large numbers capable of yielding a representative statistical sample.

The transits are generally confirmed by ground-based telescopes, and at least three events are required for a positive identification. This conservative approach means that the detection of an Earthlike planet requires at least three years of observations. But interesting results from *Kepler* came trickling in even before that. By 2011, five candidates of less than two Earth masses were found at distances from their stars comparable to Earth's. *Kepler* continued to find hot Jupiters, but it also found "puffy planets" – also known as "hot Saturns," large planets that are so tenuous they could float on water, like Saturn. The existence of super-Earths and super-Neptunes, both larger than their eponyms, shows that exoplanets span the full range of planetary sizes. But *Kepler*'s watershed finding was that small planetary masses like the Earth are common. Hot Jupiters gave way to the planets we all knew and loved.

There are two main criteria for a planet to be "Earthlike." First, it must be in what astronomers call the habitable zone, or the "Goldilocks zone," the distance from the parent star such that at least some surface water will be liquid. Life as we know it requires liquid water, so a key component of the search for this life is the search for liquid water. If the temperature is too hot, water will evaporate, and if it is too cold, it will be entirely ice (although as noted in our explorations of Europa, Enceladus, and Titan, liquid oceans underneath the surfaces of ice worlds are possible habitable zones). The Earth is the only planet in our Solar System in the habitable zone: Venus may have been there in the past, when the Sun was dimmer. But dwelling in the habitable zone is not a guarantee of habitability: we saw in Chapter 2 how a runaway greenhouse effect turned Venus into a hellish inferno.

Second, a planet must be the right mass. If it is too small, any atmosphere will escape its gravitational pull. For example, with a mass only 11% of the Earth's, Mars has lost most of its atmosphere. If a planet is too massive it will retain sufficient atmosphere to become a Neptune or even a gas giant. Near the end of this chapter we look at some even more sobering requirements that constrain the habitability of planets, some of which arise from non-astronomical considerations.

Just as planets are found in a wide range of sizes, so are stars. The stellar family encompasses sizes from the very massive, known as O- or B-stars, which are typically 5 to 20 solar masses but can be much larger, to M-type stars, which possess about a half to less than 10% of the Sun's mass. In between are A-stars, with about twice the Sun's mass; F-stars, which are just above the Sun in mass; solar-type

G-stars; and stars that are just below the Sun in mass, K-stars. The more massive a star is, the brighter and hotter it is. How big does a planet have to be before it is a star, that is, before it is sufficiently massive and hot to sustain thermonuclear reactions? Astronomers don't yet agree on an exact number, but it's in the ballpark of 10 to 30 times the mass of Jupiter (which is 318 the mass of Earth), or about 7.5% the mass of the Sun. Brown dwarfs occupy a transition zone between planets and stars: these bodies are massive enough to sustain incomplete nuclear fusion, but they do not generate sufficient heat and light to produce a clement environment for life as we know it. Planetary companions have been discovered around several brown dwarfs.

The suitability of a star to support and sustain life is determined by its lifetime, and the more massive a star, the shorter its life. Solar-type stars stick around in a stable form for about 10 billion years; our Sun is middle-aged. O-stars are a flash-in-the-pan, living for a few million years, while M-stars are the stellar Methuselahs, persisting for hundreds of billions of years. Since the Universe is about 14 billion years old, almost every M-star that formed is still in existence. This long lifetime means that they dominate the stellar population: about three quarters of stars are M-types. Early searches for Earthlike planets focused on G-type stars – and possibly F- and K-type, but astronomers now realize that M-stars not only have habitable zones, but they have the longest time for evolution to run its course. One problem with M-stars is that the habitable zone is very close to the star, which means any Earthlike planets would be tidally evolved and keep the same face toward the star, as we earlier saw is the state of most planetary moons, including our own Moon. But a heavy cloud cover might temper the differences between night and day.

After atoms formed, the Universe was mainly hydrogen with some helium and a small amount of lithium. Planets must also have sufficient heavy elements, which astronomers call "metals" (even if they aren't true metals), to sustain life. The building blocks of life – carbon, oxygen, and nitrogen – form in the interiors of stars, as do heavier elements such as silicon, magnesium, and iron that make up Earth's rocks. Our Sun is a third-generation star: the material on our planet and in our bodies was cycled through the interior of a giant star and blown out into space twice. Heavier elements, including some that are crucial to the development of technology, are formed within the expanding shells of supernovas, which are dying exploding massive stars. Heat-producing radioactive elements crucial to initiating and sustaining plate tectonics, which we saw earlier is the organizing theme of terrestrial geology, are similarly formed in these explosions. Supernovas are true cosmic birthplaces, as the expanding remnants of these dead stars also cause the collapse of interstellar gas and dust clouds to form new solar systems like our own.

Finally, failing after a lifetime of 6.5 years, when two of the four reaction wheels that sustained its stability died, *Kepler* detected a statistical sample that was beginning to look more like our Solar System. Moreover, enough exoplanets were detected to start "debiasing" the sample: that is, making estimates of the true numbers of each type of planet. For example, if we estimate from their sizes, distances from the star, and likely orbital tilts – all of which define the probability that a planet will transit a star – that about 1% of terrestrial planets at a given distance from their star will actually transit, we can just multiply the number of similar detected exoplanets within our sample of 150,000 stars times 100 to get the actual number of existing exoplanets at that mass and distance from the star. That number can be further extrapolated to all the rest of the stars in our Galaxy. This mathematical process of debiasing is applied to many fields of astronomy to estimate the true number of a class of objects that we can never observe in its entirety. By mid-2016, 2,325 confirmed exoplanets had been discovered by *Kepler*, and over 4,000 additional objects await confirmation (the vast majority – about 90% – are expected to be confirmed). Using debiasing, astronomers estimate that nearly 20% of stars have Earth-sized planets in the habitable zone: just around sunlike stars, that means there are 11 billion potentially habitable "Earths above" in our Galaxy alone. If we extrapolate the results to M-type stars, the

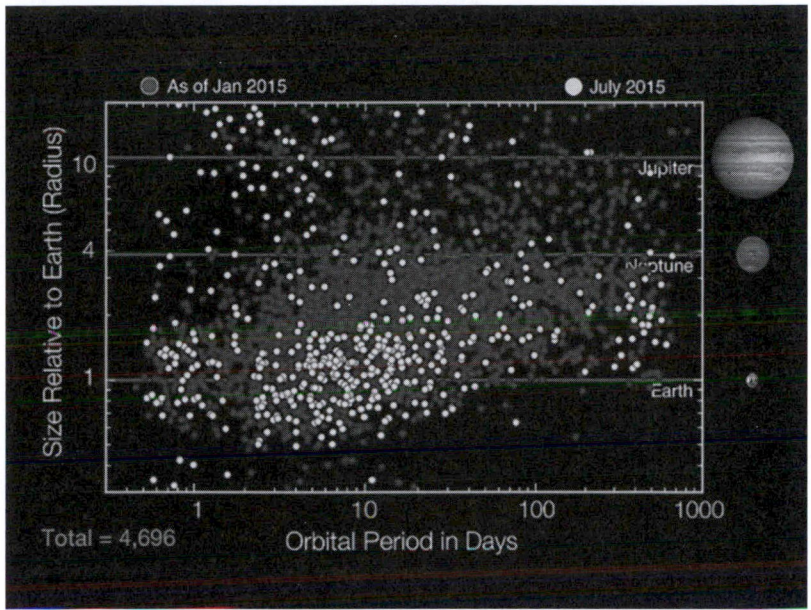

FIGURE 10.4 A summary of the exoplanet candidates detected by *Kepler* as of July, 2015; nearly half have been confirmed. NASA. Graph by W. Stenzel. See plate section for color version.

number becomes a staggering 40 billion. *Kepler*'s results imply that most stars have planets, but astronomers are still grappling with the number of planets in a typical solar system. Figure 10.4 summarizes the detections of exoplanets by *Kepler*.

While detections moved into sufficient numbers to begin inferring the statistics of exoplanets, scientists were also advancing physical characterization through both theoretical modeling and detection. Methane, water, carbon monoxide, and carbon dioxide have been discovered, although in their gaseous form. Searches are underway for rings, moons, and even volcanoes, which could be detected through gaseous emissions, especially sulfur dioxide. The similarity between terrestrial and extrasolar geology depends very sensitively on how much radioactive material – which provides the powerhouse for plate tectonics – exists in the interiors of a specific exoplanet. One possible exotic planet is the water world: a body entirely covered by a deep ocean devoid of any terra firma. Such a planet might form when an ice-coated planet migrates into the habitable zone where its vast reservoir of water melts. One exoplanet (BD+20594b) has half the diameter of Neptune but is entirely rock.

The detection of oxygen on an exoplanet would be a breakthrough, although the signature of oxygen is not necessarily a marker for life. For example, water molecules in a planetary atmosphere can be broken down into hydrogen and oxygen by high-energy radiation from its sun. Detecting the green fingerprint of photosynthesis that is the basis of most life on Earth is difficult because many minerals mimic the color of green plants. The same is true of bacteria: their spectra are in general just not sufficiently unique to identify them.

Besides exotic planets there are untethered ones, lonely worlds traveling by themselves in deep space, not in orbit about any star. These so-called rogue planets either formed on their own or were ejected from their parent star. Although only a handful of rogue planets have been confirmed, billions may exist in our Galaxy.

As milestone after milestone for finding and characterizing extrasolar systems was passed, *Kepler* reignited the interest in the elusive search for extraterrestrial intelligence. The existence of life, particularly intelligent life with a technology capable of interstellar communication, is still in the realm of speculation. During my life as a scientist, we have made great progress in detecting planets outside our Solar System, but we are no closer to finding life elsewhere, or even to clearly understanding how it arose.

A single equation that combines the uncertainties surrounding life in the Universe, and more specifically, that predicts the number of advanced technological civilizations in our own Galaxy, was developed by the radio astronomer Frank Drake of Cornell University in 1961. Along with Carl Sagan and the Soviet astrophysicist Iosif Shklofsky, Drake did key research that sparked the public's interest in what became known as SETI, the Search for Extraterrestrial Intelligence.[1] (I was privileged to have been both Frank Drake's and Carl Sagan's student during my graduate years at Cornell.) Drake's equation can be written as:[2]

$$N = N_* \times f_p \times n_e \times f_l \times f_i \times f_c \times f_L$$

where $N$ is the total number of stars in our Galaxy that support at least one planet with an advanced technological civilization, one that would be likely to contact the inhabitants of Earth; $N_*$ is the total number of stars in our Galaxy; $f_p$ is the fraction of these stars that have planets; $n_e$ is the number of planets that could sustain life as we know it around each of those stars; $f_l$ is the fraction of these planets on which life actually arises; $f_i$ is the fraction for which intelligent life arises; $f_c$ is the fraction of intelligent life that develops communications technology; and $f_L$ is the fraction of each planet's lifetime that is favored by an advanced technological civilization.

If Drake's equation looks daunting, think of the analogy of a coin toss. If you were to calculate the probability of tossing four heads in row, you would multiply the probability of getting a head (1/2) four times: $(1/2) \times (1/2) \times (1/2) \times (1/2)$ or 1/16. On average only one in 16 consecutive tosses would get you four heads in a row. Drake's equation is similar: each $f$ above, represents the probability that you would find what you are looking for. All the probabilities multiplied together yield an estimate of the probability of a communicating civilization. Multiply that number (a small fraction) times the number of Earth-like planets, and you have the number of potential communicable civilizations in the Galaxy.

The equation is "straightforward," as mathematicians like to say, but the determination of each number is difficult, and the numbers get more uncertain as we move from left to right. The number of stars in our Milky Way Galaxy is about 200 billion. This number is uncertain by a factor of plus-or-minus two: the actual number could be 100 or 400 billion. But we're fairly certain it's in that range. Many of those stars create an environment that is not conducive to sustaining life, even if they have planets: as we noted earlier, hot massive O- and B-stars burn up within tens of millions of years – not enough time for evolution to gain any traction. Other stars are unruly, spewing out gigantic flares or shedding sheets of high-energy particles and powerful radiation. But such stars are a tiny fraction of the total stellar population: A, F, G, and M stars are over 90% of the total. Thus most stars can be counted as possible havens for life. Our latest thinking about the formation of solar systems – that the central star and its associated planets condense from a collapsed cloud of gas and dust – suggests that nearly every star should have a planetary system, as *Kepler* seems to suggest. The impetus for the collapse of a cloud in a star-forming region is often an expanding supernova, which we saw is the incubator for heavy elements that are in turn supplied to the nascent solar system to provide the raw materials for life, technology, and the internal heat that drives geology. Figure 10.5 shows another transcendent *Hubble* image: the "Pillars of Creation" that are great clouds of gas and dust that are the birthplaces of stars and their planets.

*Kepler* yielded hard data on the estimate of the number of Earth-like planets, which we saw is about 40 billion. But those numbers are the more certain parts of Drake's equation – the probability that life, or that intelligent life, or that intelligent life with a technology capable of communicating evolved, is as uncertain as ever. And most uncertain is the longevity of such civilizations.

I remember sitting in classes with Drake and Sagan who both created a world of optimism in which contact with extraterrestrials was not only inevitable, but just around the corner. This belief was seen as the final extension of the Copernican Principle: there is nothing special about us or our place in the Universe. The rosy arguments went along these lines: since life arose on Earth almost as soon as it could, near the end of Late Heavy Bombardment, it should arise on all Earthlike planets. Intelligent life should arise everywhere because it seems to have arisen on Earth more than once: sea mammals such as whales and dolphins are intelligent (although they are incapable of developing a technology). A level of engineering capable of communication through space is likely to arise because at least three

FIGURE 10.5 *Hubble* image of the "Pillars of Creation," a star-forming region in the Eagle Nebula about 7,000 light years from the Earth. NASA/Space Telescope Science Institute. See plate section for color version.

civilizations developed technology: the Europeans, the Mayans, and the Chinese. M-type stars were not even considered as parent stars when Drake's equation was presented. Do their longer lifetimes mean that intelligent life will inevitably arise?

There is one note of caution centered around the longevity of civilizations. In the age of the atomic bomb, it was admitted that the very force that propelled us forward in evolution – the pressures of competition – could lead to our destruction. One corollary of the Copernican Principle, sometimes known as the Principle of Mediocrity, is that *Homo sapiens* should not be at the beginning or end of its stay on Earth, rather we should be in the middle. This thinking consigns us to a short sojourn, a mere blink in the cosmic scale of time.

In 1960, Frank Drake established the first formal SETI program: Project Ozma, which searched for regularly repeating radio signals from two nearby stars. NASA and the National Science Foundation, as well as the Soviets, maintained a modest program throughout the years. Disaster struck in 1978 when Senator William Proxmire turned his sharp eye and tongue to SETI. SETI was laughed off the Hill as Congress spoke of the search for "little green men." The program was defunded and awarded the notorious Golden Fleece Award. Sagan scrambled down to Washington and convinced Proxmire to restore funds for SETI. As technology and computer power have advanced, the search has continued, searching millions of channels with hundreds of hours of telescope time. SETI uses primarily radio frequencies, although optical and gamma-ray searches are underway. There is even a "citizen–scientist" program called SETI@home, in which anyone can use their personal computer to sift cosmic radio signals and search for patterns suggesting an intelligent source.

SETI has failed to find any evidence of extraterrestrial life. This failure, coupled with no credible evidence that we have ever been visited by extraterrestrial life, has led to Fermi's paradox, named after the Italian–American physicist Enrico Fermi who first publicized it. Put simply, if the Universe is teeming with life, why haven't we seen or heard anyone? A number of ideas have been advanced to explain the paradox: we are in a celestial zoo, consigned to observation status by intelligences far beyond ours; we haven't been looking long enough, or in the right places, or in the right ways; advanced civilizations do not care about communicating, at least with us; they are in another plane of existence, having uploaded their consciousness to computers; civilizations do destroy themselves before they can communicate; or the most sobering, that we are alone or nearly alone. Perhaps the most vexing aspect of Fermi's paradox is that it opens the door to a possible violation of the Copernican Principle. If we are alone, we – and every life form on Earth – are special, very special.

Another response to Fermi's paradox is that Drake and his followers may be all wrong. Perhaps the Universe isn't teeming with intelligent life, even if it is teeming with life.

In their book *Rare Earth*, Peter Ward and Don Brownlee of the University of Washington outline a series of considerations that render intelligent life nearly impossible.[3] Most problems either center around the inhospitability of the Universe, or around the unlikely series of events and situations that coddled life on Earth to the state in which it is now. First, Ward and Brownlee temper the idea of the habitable zone: even the Earth itself may have gone through a snowball stage in which its entire surface was ice, and we already saw that many Earthlike planets might be water worlds. The Earth's Moon – which may itself be a rare type of satellite unusually large in comparison with its primary – counterbalances the tilt of the Earth, which would otherwise lead to wild climate fluctuations. Jupiter clears the Solar System of the lethal impactors, described in Chapter 4, which could also transport additional water to a habitable planet to make a water world even more likely. Plate tectonics, which depends on the presence of heat-providing nuclear isotopes in a planet's interior to power it, is possible only in a later generation star. This driving principle of Earth's geology commands a leading role in the trajectory of life by sequestering the greenhouse gas carbon dioxide (see Chapter 2), and by causing mountains to form. Shallow seas form from the washed-down debris shed by these mountains to provide a locus for sea life to walk onto dry land. Life on Earth existed in a simple, single-cell form until the Cambrian explosion, about 542 million years ago, when complex life arose in a relatively short period of time.

Rare Earth adherents suggest that pre-Cambrian life is what is normal and that we are freaks.

We still don't know how life arose, let alone intelligent life, or even complex life. We don't know anything about life as we don't know it. For life as we know it, we only have a sample of one. For planets, we had a sample of eight or nine and, as every planetary encounter showed, each world was different than any we could imagine.

The Rare Earth hypothesis leads to a serious violation of the Copernican Principle that there is nothing special about us or about our place. There is no easy way out of this paradox, but there are some interesting ideas from the field of theoretical physics. Scientists have long noted that the physical constants that guide the Universe seem fine-tuned to be hospitable to life, as if the Universe was created just for us. The strength of the gravitational force leads to closed planetary orbits; there exists the right number of physical dimensions; the ratio of the various forces of nature leads to stable atoms, etc. Of course, life will arise to be compatible with its own world, so this idea is somewhat tautological. Another type of life may have arisen with other constants. Our compatibility with the present Universe, which is known as the anthropic principle, has led physicists to hypothesize an infinity of universes – multiverses – each with its own set of fundamental constants, and each giving rise to its own form of life. Thus the smallness of our dwelling within this Universe is paled by the presence of an infinity of universes, each with its own life forms. The idea of multiverses is the next step in the Copernican view: just as the Earth, the Solar System, and our Galaxy are not special but one of billions, perhaps our Universe is one of an infinite number. But there is a difference with this last step: experimental science peeled away layer after layer to demolish our specialness, but these multiverses are beyond what we can observe with our current technology and knowledge. They remain speculative, as science can only be built on experimental evidence.

Indeed, the Teacher of the Dalai Lama was right: there are more worlds than I could ever conceive of.

NOTES

1 Shklofsky, I. and Sagan, C. 1966. *Intelligent Life in the Universe.* New York: Dell.

2 Sagan, C. *Cosmos.* 1980. New York: Random House.

3 Ward, P. and Brownlee, D. 2000. *Rare Earth.* Gottingen: Copernicus.

# Epilogue

Whenever I give a public talk on space exploration, or speak with someone casually, such as on an airplane or on a bus, I am often asked why NASA isn't doing more: "Why don't we have bases on the Moon? Why aren't there astronauts on Mars, and why didn't *New Horizons* go into orbit around Pluto or send back a sample?" My answer is invariably, "we would like to do all those things, if only NASA had the funds. We can think of all sorts of marvelous missions, but our budget is so constrained. We are limited to just a few high-priority projects."

In the past we were able to do so much more, but as NASA's funding dwindled from 4.5% of the national budget during the peak years of the *Apollo* lunar program, to less than 0.5% today, we are confined to a bare-bones portfolio of exploration. The dreams of my childhood, and of so many others, have been diminished as the budget for scientific work in general has declined. It is not only NASA. IBM also used to have an outstanding group of researchers, and of course there was Bell Laboratories, which spawned so many Nobel Prizes, including the work of Arno Penzias and Robert Wilson with their discovery in 1964 of the remnant thermal radiation from the Big Bang, the Universe's moment of creation almost 14 billion years ago. Less commonly someone will say "Why spend all that money out in space?" I will respond: "None of the money is spent in space. It is all spent here on Earth providing good jobs that cannot be outsourced or eliminated."

*New Horizons* will continue on to the smallish Kuiper Belt Object 2014 MU69, Juno started exploring the interior of Jupiter in mid-2016, the *Mars Atmosphere and Volatile EvolutioN Mission* (*MAVEN*) is orbiting Mars to understand how it lost its atmosphere, and the *InSight Lander* will explore the interior of Mars with seis-mometers and heat-flow detectors. *Mars 2020* will follow on the heels of the *Mars Curiosity Lander* to explore possible habitable environ-ments on Mars and to search for life. But it will only cache a collection of samples for later return because we don't have the funds – or the technological resources – to return this valuable cargo in the near future. A mission to Europa is stumbling along in fits and starts – a payload of scientific instruments was selected in 2015, but it won't launch before 2022. Small bodies are on the docket too. Beginning in 2018, *OSIRIS-REx* will study a primitive asteroid and return a sample to Earth in 2023, three years after the Japanese *Hayabusa 2* mission will do the same, landing on the tiny carbon-rich asteroid 162173 Ryugu, which has a surface area one-seventh that of Manhattan.

NASA hasn't been to Venus in decades (although ESA's *Venus Express* spent the last decade exploring the Queen of Heaven in detail), and there are no funded plans to explore Earthlike Titan or the crin-kles and fountains of Enceladus. *New Horizons* revealed unexpected activity in the Kuiper Belt: perhaps this collection of space junk is instead a treasure trove of protoplanets all in different stages of evolu-tion. Eris is a very bright KBO that rivals Pluto in size: it is as dazzling as fresh snow on the coldest day. Does it have glaciers too, or active ice plumes like Enceladus? Is it an ocean world with subsurface alien seas at the edge of the Solar System? We could answer these ques-tions and more if we had funding equivalent to the *Apollo* days; at the same time we would be turning out a well-trained core of scientists, engineers, and technical experts.

The European Space Agency (ESA) also has a program of explo-ration. In many ways ESA's plans are more stable than NASA's because they are agreed upon by a consortium of nations and thus less subject to the vagaries of any one political system. The *Rosetta* mission's lander *Philae* was the first craft to land on a comet for detailed scrutiny, and the mother ship performed a controlled crash on the comet's surface on September 30, 2016. ESA has a solid plan to study Jupiter and its moons with its *JUpiter ICy moons Explorer*

(*JUICE*) mission planned for launch in 2022. China has orbited and landed on the Moon with its *Chang'e* program, and India's *Chandrayaan* program has sent spacecraft to orbit and characterize the moon.

Once NASA had a plan to establish a human colony on Mars by now: these plans have been eternally postponed. Of course, the golden days of space exploration took place within the landscape of the Cold War: our goal was to get a crewed mission to the Moon before the Soviets, and to prevent the sovietization of space. Will it take another Space Race, perhaps with the Chinese or Indians, to reboot NASA's space program?

War and international competition drive technology, which in turn drives science. But the true payoffs are in the civilian arena. Challenging technical programs produce great human capital. After the Manhattan Project, which produced the atomic bombs that leveled Hiroshima and Nagasaki and killed upwards of 200,000 civilians and soldiers, the graduates of this immense military enterprise went on to continue the golden age in physics that had begun at the turn of the last century with the discovery of relativity and quantum mechanics. These supreme discoveries in physics led to transistors, then integrated circuits, then the tech boom. You never know where science will lead.

Science was once a great part of our cultural conversation, and politicians panicked whenever an enemy nation's students showed superiority in math and science. The worth of science and technology was never questioned, and everyone understood the value of a technical education, even for non-scientists. Scientific knowledge enhances critical, analytical thinking and underscores the virtues of skepticism and empiricism. Scientists and the whole enterprise of science had authority. Now scientists are seen as a consortium of special interest groups, each begging Congress to bankroll their pet projects. This perception of science, and more profoundly the thinking that drives it, is dead wrong.

The golden age of space exploration produced a corps of trained scientists and engineers; the smart and industrious children who came of age in the 1950s and 1960s were drawn to science, technology, engineering, and math, areas we now call collectively STEM. Until recently the United States had a stock of trained STEM people, and they could all find jobs. As my Dad would say with his inimitable bluntness: "you'll never find an engineer walking the streets." And they were not only college graduates, but the highly skilled machinists, welders, drafters, etc. who provide the groundwork for leadership in technology and manufacturing. As the students of the post-Sputnik generation have entered more non-technical fields, or clung to closely defined vocational majors, we lack the type of profoundly educated, agile workforce that technological progress demands. The lure of space has always been a "hook" to encourage students to major in basic STEM fields that provide that expertise and versatility. That lure is no longer so visible.

Have the physical sciences simply been eclipsed by the biological sciences? Is the current century the century of biology, just as the twentieth century was the century of physics? With climate change and its associated ecological disasters; with a sixth great extinction unfolding; and with medical promises of curing diseases knocking at our door, shouldn't we prioritize the life sciences?

Science cannot be separated into categories like that. The structure of DNA was discovered from data collected by the physicist Rosalind Franklin, using classical techniques of crystallography. Advances in medical science were aided by the development of image processing techniques perfected by space scientists. Miniaturization of electronics, robotics (you can't easily go out into space to fix anything) were all generated by the Space Age. If you simply delete a core area of science, you don't know what future discoveries will be thrown out. Perhaps the next great discovery in environmental science – say, a way to sequester carbon dioxide – will come from an area of physics or chemistry. A program to deflect a killer asteroid will encompass a vast array of scientific and engineering disciplines, from classical celestial dynamics, to mechanical engineering, to observational astronomy. An advanced society simply needs a cohort of trained scientists and engineers in every field or it is deficient.

In the US Senate Science Committee in March, 2012, astronomer and director of the Hayden Planetarium, Neil deGrasse Tyson, testified that "Right now, NASA's annual budget is half a penny on your tax dollar. For twice that – a penny on a dollar – we can transform the country from a sullen, dispirited nation, weary of economic struggle, to one where it has reclaimed its 20th century birthright to dream of tomorrow."

# Glossary

**aphelion** – the farthest point in an object's orbit around the Sun.

**apparition** – an appearance of a celestial body in the sky.

**aqueous** – pertaining to water, such as aqueous alteration.

**arc second** – 1/3600 of an angular degree; an arc minute is 1/60 of a degree.

**asteroid** – one of the small sub-planet sized bodies of the Solar System orbiting mainly between Mars and Jupiter.

**astronomical unit (AU)** – the distance between the Earth and the Sun; 150 million kilometers or 93 million miles.

**blue-shift** – a decrease in wavelength of an approaching body's spectrum.

**bolide** – a large meteor.

**caldera** – a depression on large volcanoes caused by the outflow of magma.

**chaotic rotation** – a state in which a celestial body exhibits no regular rotation period.

**charge coupled device (CCD)** – a solid-state electronic camera widely used as an astronomical detector.

**comet** – a small body containing volatiles that outgas as it approaches the Sun.

**conjunction** – the apparent close placement of objects, usually planets or spacecraft, in the sky.

**cryovolcanism** – volcanic processes involving ice rather than molten rock.

**deuterium** – heavy hydrogen consisting of a neutron and a proton in the nucleus of the atom plus an orbiting electron; regular hydrogen consists of just a proton in the nucleus and an electron.

**Doppler effect** – the change in wavelength of the spectrum of a body moving with respect to an observer.

**eclipse** – the blockage of light from a celestial body by another body.

**ecliptic** – the path of the Sun and the planets on the sky.

**ellipse** – the oval shape that defines the trajectories of objects orbiting the Sun; mathematically it is the intersection of an inclined plane with a cone.

**elongation** – the angle between the Sun and a planet as seen from the Earth; commonly it refers to the maximum of this angle, when the planet appears the farthest from the Sun.

**evaporites** – minerals formed from the evaporation of water or other liquids; they include carbonates and sulfates.

**exoplanet** – a planet that orbits a star other than the Sun.

**extremophiles** – organisms that can live in inclement environments such as extreme hot, cold, salinity, acidity, depth, or altitude.

**gaseous planets** – planets with a composition mainly of hydrogen and helium; in our Solar System they are Jupiter, Saturn, Uranus, and Neptune.

**geocentric** – a cosmic system with the Earth at the center and all other bodies orbiting it.

**greenhouse effect** – the trapping of infrared radiation by certain gases to cause a warming effect.

**habitable zone** – an area where life can exist; food, energy, and liquid water are the prime requirements.

**heliocentric** – a cosmic system with the Sun at the center with the planets orbiting around it.

**hematite** – an iron-bearing mineral that is often formed from standing water.

**hydrocarbons** – the simplest organic molecules consisting of hydrogen and carbon; when expanded to include the building blocks of life, oxygen and nitrogen are present as well.

**JPL (Jet Propulsion Laboratory)** – NASA's premier center for the robotic exploration of deep space, operated by the California Institute of Technology.

**Kepler's laws of planetary motion** – (1) the planets move in ellipses about the Sun; (2) the orbits of the planets sweep out equal areas in equal times, so that objects orbit faster the closer they are to the Sun; and (3) the square of a planet's orbital period divided by its average distance to the Sun cubed is a constant.

**Kuiper Belt Objects (KBOs)** – sub planet-sized bodies (also known as Trans-Neptunian Objects, or TNOs) in orbit about the Sun beyond the orbit of Neptune.

**Late Heavy Bombardment** – a time of an unusually high number of impact events after the formation of the planets, between approximately 4.1 to 3.8 billion years ago, and tapering off after that.

**light curve (or lightcurve)** – The brightness of a celestial body with time: for example during its rotation.

**light year** – the distance light travels in one year; about 6 trillion miles or 63,241 astronomical units.

**lineaments** – a linear feature that is an expression of an underlying geologic structure.

**mare (pl. maria)** – smooth, basalt-rich areas of the Moon that are believed to be vast areas of melt formed by impacts during the period of the Late Heavy Bombardment.

**mean motion resonance** – an orbital configuration between one or more bodies in which one body's period is a simple multiple of the other(s), leading to frequent alignments.

**meteor** – a fragment of an asteroid, comet, or meteoroid as it burns up in the Earth's atmosphere; also commonly known as a shooting star.

**meteorite** – a fragment of an asteroid or comet that has hit the Earth's surface.

**meteoroid** – asteroids smaller than one meter (39 inches).

**methanogens** – bacteria that produce methane and do not require oxygen for metabolism.

**near-Earth objects (NEOs)** – asteroids or comets that come to within 1.3 AU of the Earth.

**Newton's laws of motion** – (1) an object at rest stays at rest; (2) the force on an object equals its mass times the acceleration of the force; and (3) for every action there is an equal and opposite reaction.

**nodes** – the point at which two orbits cross.

**Occam's razor** – the principle that the simplest explanation is usually the correct one.

**occultation** – the passage of an astronomical body in front of a more distant object.

**Oort Cloud** – a massive spherical cloud of billions of comets about 50,000 AU from the Sun.

**perihelion** – the closest point in an object's orbit around the Sun.

**planetesimals** – solid objects about a kilometer (0.6 miles) in size that were the building blocks of planets.

**plasma** – electrically charged gas.

**plate tectonics** – the organizing principal of terrestrial geology in which the Earth's surface is composed of a number of rigid plates that ride on a semi-liquid crust.

**polymer** – a large molecule consisting of repeated components.

**precession** – a change in the axis of rotation of a planet or satellite.

**prograde** – the usual sense of rotation or orbital motion for a planet, counterclockwise as observed from above the Solar System.

**protoplanet** – an embryonic planet that has undergone differentiation into a core, mantle, and crust.

**red-shift** – an increase in wavelength of a receding body's spectrum.

**refractory materials** – components such as metals that have very high melting points; they were the first to condense when the Solar System formed.

**retrograde** – rotation or orbital motion opposite to that of prograde, or clockwise as observed from above the Solar System.

**scarp** – a steep slope usually due to a geologic fault.

**shield volcano** – a large, wide volcano formed from successive lava flows.

**solar wind** – the supersonic charged particles that are ejected from the Sun and are entrained in its magnetic field to permeate the Solar System.

**spectrum (pl. spectra)** – a measure of the intensity of radiation emitted or reflected by a body as a function of wavelength.

**synchronous rotation** – a dynamical state caused by tidal evolution in which an object orbiting a larger body keeps the same hemisphere toward the larger body.

**terrestrial planets (or bodies)** – the smaller, rocky planets of the Solar System: Mercury, Venus, Earth, and Mars.

**third zone** – the region of small icy bodies outside the orbit of Neptune, comprising a main category of planetary bodies in addition to the terrestrial and gaseous planets.

**transit** – the passage of a celestial body in front of a larger body.

**Trojan asteroids** – the group of asteroids trapped in the gravitationally stable zones of Jupiter's orbit.

**volatiles** – gases.

**wavelength** – the distance between crests of the waves that comprise all electro-magnetic radiation, including light, radio, and infrared radiation.

# Index